ROUTLEDGE LIBRARY EDI
URBAN PLANNING

Volume 20

COST-BENEFIT ANALYSIS IN
URBAN & REGIONAL PLANNING

COST-BENEFIT ANALYSIS IN URBAN & REGIONAL PLANNING

J. SCHOFIELD

Routledge
Taylor & Francis Group

LONDON AND NEW YORK

First published in 1987 by Allen & Unwin

This edition first published in 2018
by Routledge
2 Park Square, Milton Park, Abingdon, Oxon OX14 4RN

and by Routledge
711 Third Avenue, New York, NY 10017

Routledge is an imprint of the Taylor & Francis Group, an informa business

© 1987 J. A. Schofield

British Library Cataloguing in Publication Data
A catalogue record for this book is available from the British Library

ISBN: 978-1-138-49611-8 (Set)
ISBN: 978-1-351-02214-9 (Set) (ebk)
ISBN: 978-1-138-49452-7 (Volume 20) (hbk)
ISBN: 978-1-138-49454-1 (Volume 20) (pbk)
ISBN: 978-1-351-02602-4 (Volume 20) (ebk)

Publisher's Note
The publisher has gone to great lengths to ensure the quality of this reprint but points out that some imperfections in the original copies may be apparent.

Disclaimer
The publisher has made every effort to trace copyright holders and would welcome correspondence from those they have been unable to trace.

COST-BENEFIT ANALYSIS IN URBAN & REGIONAL PLANNING

J. SCHOFIELD

Department of Economics
University of Victoria, British Columbia

London
ALLEN & UNWIN
Boston   Sydney   Wellington

Allen & Unwin, the academic imprint of
Unwin Hyman Ltd
PO Box 18, Park Lane, Hemel Hempstead, Herts HP2 4TE, UK
40 Museum Street, London WC1A 1LU, UK
37/39 Queen Elizabeth Street, London SE1 2QB, UK

Allen & Unwin Inc.,
8 Winchester Place, Winchester, Mass. 01890, USA

Allen & Unwin (Australia) Ltd,
8 Napier Street, North Sydney, NSW 2060, Australia

Allen & Unwin (New Zealand) Ltd in association with the
Port Nicholson Press Ltd,
60 Cambridge Terrace, Wellington, New Zealand

First published in 1987

British Library Cataloguing in Publication Data

Schofield, John A.
  Cost–benefit analysis in urban and
  regional planning.
  1. Regional planning——Cost effectiveness
  2. City planning——Cost effectiveness
  I. Title
  338.4'336161   HT391
  ISBN 0–04–338145–6

Library of Congress Cataloging in Publication Data

(Data applied for)

Typeset in 10 on 12 point Times by Columns of Reading
and printed in Great Britain by
Biddles of Guildford

For Josie

Preface

The cost–benefit method has received relatively little attention in the literature on techniques of urban and regional planning. The aim of the current book is to help to fill this void. It outlines the principles of cost–benefit analysis (CBA) in the context of urban and regional studies, and indicates the usefulness and limitations of the technique by reference to examples of its implementation as an instrument of urban and regional planning.

The book is designed to appeal to upper-level undergraduate students in economics, geography, and town and country planning, and to graduate students with an interest in evaluation techniques in these disciplines and related applied social sciences such as public administration, transportation and health planning. It is also designed for practitioners of CBA from these or other backgrounds, and professionals with a basic knowledge of economics who are involved in different areas of urban and regional planning. The presentation is at the intermediate level using predominantly graphical methods of analysis, so that no more than a foundation course in the principles of economics is required by way of preparation.

The theory of CBA is developed in Part I. While examples are used in Part I to illustrate the empirical relevance of the principles, detailed discussion of applications is left to Part II. Part II shows how the method can be used to address a variety of issues in urban and regional planning. The term 'urban and regional planning' is taken to refer to the process of attempting to optimize the use of resources in areas of public policy that have a subnational dimension, either because decisions are made at the subnational level or because national policy is directed at subnational units. Thus applications of CBA in planning the provision of local health and social services and in evaluating local or regional economic development efforts are included in the discussion, along with examples of application concerning the land use issues of physical or town and country planning.

I am indebted to the University of Victoria for providing a sabbatical leave during which a substantial part of the manuscript was prepared. I am also indebted to students who have helped to refine my presentation of the principles of CBA over the years, to my colleague, Dr Gerry Walter, for discussion of various matters included in the text and to Dr Jeremy Alden for reading the entire book in draft form and for making helpful comments and suggestions. The usual disclaimer applies.

Finally, I wish to thank Michele Armstrong, Rachel Decaria, Mavis

Murray, Anna Petrie, Priscilla Shiu and my wife (whose own research was interrupted in the process) for undertaking the typing at different stages of the book's evolution.

J. A. Schofield

Contents

List of tables

List of figures

COST-BENEFIT ANALYSIS IN URBAN & REGIONAL PLANNING

1
Introduction

To date, the cost–benefit method has been relatively neglected in surveys of the techniques of analysis used in urban and regional planning. There exists no detailed assessment of the method such as provided by Isard *et al.* (1960, 1972) for a variety of other techniques, by Richardson (1972) for input–output analysis or by Adams and Glickman (1980) for regional econometric modelling. Indeed, some early surveys of analytical methods for planning at the subnational level omit reference to cost–benefit analysis (CBA) altogether.[1]

CBA is a technique for assisting with decisions about the use of society's scarce resources. Specific questions which the method helps to address are: Is a project or programme worthwhile? What is its optimal scale of operation? What is the optimal timing of its initiation? What is the relative merit of different projects or programmes? There will be no claim in this book that definitive answers to these questions will always derive from CBA. Given the many problems of implementation, the need for the exercise of executive judgement remains. But to the extent that careful CBA helps to narrow the area of ignorance regarding a particular issue, it can facilitate the making of sounder decisions about resource use than would otherwise be made.

In this introductory chapter, we discuss what is meant by the terms 'benefits' and 'costs', with particular reference to the distinction between purely financial benefits and costs and others identified in CBA. We also indicate the scope for empirical application of the method before outlining the plan of the book.

1.1 DEFINITION OF BENEFITS AND COSTS

In general terms, benefits are defined as contributions towards, and costs as detractions from, project or programme objectives. Thus, if the objective is a purely financial one, as profit maximization is in the private sector, an analysis comprises merely revenues and expenditures as valued in the market. Agencies undertaking initiatives in the public sector are likely to be interested in the purely financial effects of projects or programmes as these impinge upon their budgets. It may also be appropriate to conduct a purely financial analysis from the point of view of the impact on net worth for particular groups affected by policies. Examples might include migrants involved in labour relocation schemes, students in educational programmes or patients in health care programmes. Accordingly, examples of financial cost–benefit analysis in the field of urban and regional planning are included in the discussion in Part II of the book.

But CBA goes beyond mere financial appraisal to include evaluation from the wider point of view of the economy or society in general. In this context it is common to think in terms of two broad objectives: economic efficiency in the use of the resources available to society (or its dynamic analogue, economic growth), and equity in the distribution of welfare between different groups (e.g. income classes, regions, generations) within society. Analyses which focus solely on economic efficiency have come to be termed 'economic' CBAs, while those which, in addition, incorporate consideration of distributional effects are nowadays termed 'social' CBAs (Irvin 1978).

1.1.1 Economic analysis

Economic (or efficiency) effects consist of positive and negative impacts on the production and consumption opportunities, and hence utility or welfare level, of society. These are termed 'real' effects. Benefits stem from resource allocations which enhance the efficiency with which resources are converted into welfare levels; costs comprise 'opportunity costs', the opportunities forgone or welfare sacrificed as a result of diverting resources from other uses to the one(s) under analysis. Finally, in economic analysis, benefits and costs are summed over all parties affected to give aggregate measures of benefit and cost. The distribution of benefits and costs between different parties is not regarded as material.

From these definitions of economic benefits and costs it follows that economic analysis differs from purely financial analysis in four respects. First, while excluded from the narrow perspective of financial analysis, external effects on third parties are included in economic analysis. Indeed, as we show in Chapter 2, a primary purpose of CBA is to evaluate externalities, as these are not internalized in private flows of benefit and cost. For example, economic analysis of a public sector programme includes welfare impacts on groups who are neither users of the service provided nor responsible for funding the project.

Secondly, welfare impacts not priced in the market are nonetheless taken into account in economic CBA. Thus benefits include changes in consumers' surplus, that is, the excess which consumers would be *willing* to pay over and above what they *have* to pay for a commodity. Benefits also include changes in producers' surplus or economic rent – payment to factors of production in excess of supply price – as this represents a welfare gain to factors. Questions regarding measurement of consumers' and producers' surpluses are addressed in Chapter 4. In addition, intangible effects (e.g. amenity or cultural values, the value of life or time), excluded from purely financial analyses because they are not priced in the market, are included as appropriate in economic analysis. The reason for inclusion of intangibles is that protection of such things as environmental amenities, cultural heritage or life impinges on consumption opportunities and hence welfare levels just as the purchase of marketed goods and services does. The issues of principle involved in measuring these intangible effects are discussed in Chapter 5.

A third distinction between economic and financial analysis is that in economic CBA which deals with aggregate efficiency, no matter who derives the benefits or incurs the costs in the final count, 'pecuniary' effects cancel out as transfers between parties. 'Pecuniary' effects are those which benefit one party but hurt another. Unless we choose to weight the interests of the various parties differently (which we would do only in a social CBA), 'pecuniary' effects represent a mere redistribution of welfare within society, rather than a net change in aggregate welfare. As an example, transfer payments from governments to individuals are benefits to recipients but costs to taxpayers, and are self-cancelling in an economic CBA. Thus the economic benefits of a job-creation proposal exclude savings in unemployment compensation. Of course, in the context of a purely financial analysis conducted from the perspective of either the government or the particular group of workers affected by the proposal, changes in unemployment compensation are included as benefits or costs. But in economic CBA, only 'real' effects are counted.

Finally, in so far as benefits and costs are designed to reflect true economic impact, they are defined in terms of so-called shadow (or accounting) prices, which we shall refer to as efficiency prices. If observed market prices are not considered properly to reflect real additions to, or deductions from, economic welfare (as a result of market disequilibria or distortions), shadow prices are substituted for that purpose. The principles of efficiency pricing in economic CBA are explored in Chapter 5.

1.1.2 Social analysis

Social analysis addresses the issue of distribution as well as economic efficiency. At one level, this consists of the display of results on a disaggregated basis according to welfare impacts on different groups within society. At another level, social analysis extends the principle of shadow pricing to what is termed social (as opposed to efficiency) pricing (Squire & van der Tak 1975, Irvin 1978). This procedure involves the attachment of differential weights to monetary benefits and costs for different groups, to reflect judgements concerning the relative 'deservingness' of each group. Thus, benefits for disadvantaged groups are weighted, or shadow priced, more highly than benefits for other groups.

Social pricing may also involve differential weighting of benefits according to their distribution between savings and consumption. For example, benefits that are saved (and hence become available for investment) may be weighted more highly than benefits that are consumed. Such may be the case if the government wishes to promote faster economic growth, but encounters administrative or political constraints regarding use of the tax system to create additional savings at the expense of current consumption. This is a problem accepted as occurring more usually in lesser developed than in developed economies. By differential weighting, projects yielding benefits used for saving are preferred, *ceteris paribus*, to projects yielding consumption benefits.[2] The principles of social pricing, together with problems of weight determination, are discussed in Chapter 6.

1.1.3 Additional considerations

Further considerations relating to the definition of benefits and costs concern the viewpoint of the analysis, the so-called 'with-and-without' principle and the issue of double-counting. It should go without saying that the definition of benefits and costs varies with the jurisdictional viewpoint from which an analysis is conducted. Moreover, as indicated earlier, benefits and costs from the purely financial viewpoint of government or of particular groups do not coincide with those defined from the economic or broad social points of view. It is obviously important to distinguish clearly the various viewpoints from which it may be relevant to analyse an undertaking. A programme funded at the local government, provincial or state level, for example, is typically analysed from the viewpoint of the local authority, province or state economy, and often as well from the financial viewpoint of the local, provincial or state treasury. The proposal may also be analysed from national economic and disaggregated social viewpoints if these are considered to be worth highlighting. But it is essential to keep the separate perspectives distinct.

It is also obvious that in identifying benefits and costs, only incremental effects associated with the undertaking require to be counted. What matter are effects which are truly caused by, and would not occur in the absence of, the undertaking; not the same distinction, it is to be noted, as the simple difference between conditions prevailing before and after implementation of the project or programme. Identification of incremental effects involves establishment of what would have occurred in the absence of the proposal (the counterfactual condition), a hypothetical state which is never easy to define. In Part II of the book, we examine approaches to the estimation of the counterfactual condition in the context of urban and regional analysis.

Lastly, it is important in economic CBA to avoid double-counting 'real' benefits and costs, a problem which may occur more easily than is often recognized. Three examples discussed by Mishan (1982a, ch. 12) may suffice to illustrate this point. First, it is inappropriate in the case of an improved transportation link to count both the benefits of the scheme to travellers and any increase in property prices which may occur as a result of it. The latter merely reflects the capitalization of the former and represents an alternative measure of the benefits to the extent that traveller benefits are transferred to property owners in higher prices. Secondly, it would also be inappropriate to count both the user benefits of the improved link and any cost savings involved in the replacement of an existing link. The reason is that user benefits are measured, as we shall see, in terms of willingness-to-pay for the improved link or compensation required to forgo it; and these measures of benefit already take into account considerations of cost savings. Thirdly, if an irrigation scheme reduces the cost of crop production, the savings in resource costs are not to be double-counted by including as benefits any of the savings passed on from farmers to marketing agents as increased profits. In short, there are often alternative ways of defining the same benefits or costs and, in measuring the economic effects of an undertaking, it is essential to choose just one.

1.2 SCOPE OF CBA

As far as the scope of CBA is concerned, the method applies at a variety of levels of analysis and in a wide selection of fields. As indicated above, the analysis may adopt financial, economic or social viewpoints. It may also be conducted from different jurisdictional perspectives. Furthermore, the analysis may apply to either individual projects or broader programmes or policies, although, given the difficulty of measuring widespread general equilibrium effects, the conventional wisdom is that it is more safely applied the smaller the undertaking under investigation (see Prest & Turvey 1965). In addition, the analysis may be conducted before the decision is made (the *ex ante* level of analysis) or after the decision has been made as a check on performance (the *ex post* level of analysis). While the *ex ante* perspective is commonly regarded as the more valuable for planning purposes, the *ex post* approach is gaining increasing attention as a useful method for deriving lessons for the future from the past.[3]

Concerning fields of application, most of the seminal work in CBA was done in the area of water resources planning (e.g. Krutilla & Eckstein 1958, Eckstein 1961b). Over the past 25 years, however, the method has found widespread use in such areas as human resources planning (education and training, health and welfare, migration), transportation, urban renewal and housing, law and order, recreation, research and development, defence, town planning and regional development. A number of these fields of application receive attention in this book.

1.3 PLAN OF THE BOOK

The plan of the book is as follows. Part I outlines the principles of CBA, relating them to concerns which arise in urban and regional planning. Detailed discussion of applications is reserved for Part II.

Part I begins with a review of welfare economics, the theorems of which provide the conceptual foundation for the cost–benefit approach. Out of this material comes the justification for state intervention in the market and the consequent case for the use of CBA as an instrument for assisting with resource allocation decisions when markets cannot be relied upon to make such decisions optimally. The principles of welfare economics also provide a set of decision rules which constitute a framework for the implementation of CBA. For readers not interested in the details of this background material, it is possible to go straight to Chapter 3, although concepts discussed in Chapter 2 are referred to at later points in the text.

In Chapter 3 we develop the principles of investment appraisal which are required to determine whether an undertaking is acceptable, its rank in the order of alternative undertakings and its optimal size or timing. Issues relating to the evaluation, or measurement, of benefits and costs in economic and social analysis then follow in Chapters 4–8. These chapters focus, in turn, on the measurement of consumer and producer welfare surpluses, the relevance of efficiency pricing, the treatment of non-efficiency (or distributional) effects, the incorporation of uncertainty and risk into analyses, and the establishment of the interest (or discount) rate as this is used to convert the benefits and costs occurring at different points in time to commensurable values.

Part II reviews applications of CBA to selected issues of urban and regional planning. The field of urban and regional planning is seen as being concerned with more than just issues of land-use (town and country) planning; it is defined to encompass provision of all public services at the subnational level as well as consideration of the impact of projects and policies on localities or regions. Chapters 9–15 offer different perspectives on actual techniques and are devoted respectively to the issues of residential urban renewal, transportation, recreation, comprehensive land-use planning, local health and personal social services, urban and regional development implications of capital invest-ment projects, and regional policy evaluation. Of these, the first four (Chs 9–12) deal with issues of land-use planning and the last two (Chs 14 and 15) with issues of economic development at the subnational level. Chapter 16 discusses matters which arise in connection with the use of

CBA for urban and regional planning, no less than for other purposes, in the particular context of lesser developed countries. In each of these chapters an attempt is made to highlight the strengths and weaknesses of empirical work and to indicate the scope for improvement and further research. Finally, in Chapter 17, the lessons of both parts of the book are drawn together in an evaluation of the merits of CBA in general and, specifically, as an instrument of urban and regional analysis.

NOTES

1 Examples include Isard *et al*. 1960, 1972, Hall 1970, Bendavid 1972 and Masser 1972.
2 The method of social pricing may further be used to take account of concerns other than distributional matters when such concerns are not incorporated into analyses by means of monetary measures. The net economic benefits of projects that create unmeasured advantages such as protection of the environment or strengthening of national defence may be weighted more heavily than net benefits of projects that are less advantageous in these respects.
3 A forceful statement of the case for *ex post* analysis, along with empirical examples in the water resources field, is found in Haveman 1972.

PART I

Principles of cost–benefit analysis

2
Welfare economics and the foundations of CBA

The rationale for CBA is found in the propositions of welfare economics, that branch of the dismal science of economics in which we outline the necessary conditions of social welfare maximization, the problems of achieving the optimum state and the rules available for guiding decision makers on the question of whether a policy initiative moves society in the direction of improved social welfare. CBA is nothing more than a technique for assisting decision makers with this last question in situations where the public sector is called upon to provide goods and services or to assess the outcome of market interactions.

In this chapter, then, we develop the foundations of CBA as they exist in the theorems of welfare maximization. First, we define the necessary conditions of welfare maximization (or Pareto optimality) and the situations in which even a perfectly competitive market fails to promote optimality, creating thereby a need for government intervention and a consequent case for the use of CBA. Then we examine alternative rules proposed for the guidance of public decision makers in their efforts to promote social welfare through the use of CBA. As indicated in the introductory chapter, the material in .this chapter may be omitted by those who prefer to go straight to the nuts and bolts of CBA, although certain concepts introduced here are referred to at later stages in the text.

2.1 PARETO OPTIMALITY

In order to select projects or programmes which enhance social welfare, it is necessary to understand what constitutes the ideal situation at which decision makers should presumably aim, at least to the extent of moving society in the right direction. It is conventional to define this ideal as a situation in which nobody can be made better off without making someone else worse off. This is the Pareto optimum, named after the 19th-century Italian social scientist who identified this notion of maximum efficiency in the use of society's scarce resources. If we ignore for the moment the distribution of the fruits of economic activity, who gains and who loses in the process of achieving the Pareto optimum, maximization of efficiency can be viewed as tantamount to maximization of social welfare.

In order to understand the purpose of CBA, it is important to define the necessary conditions of Pareto optimality and to show under what conditions markets fail to generate these conditions. In the event of market failure there occurs a case for government intervention and the use of CBA for analysing public sector initiatives or evaluating the economic and social implications of private decisions. In this section we define the three conditions of Pareto optimality (or economic efficiency): efficiency in exchange, efficiency in production and efficiency in both exchange and production.

Efficiency in exchange, the optimal allocation of commodities among consumers, requires that the marginal rate of substitution (MRS) between commodities be the same for all consumers. MRS is the rate at which the consumer is willing to trade off (substitute) one additional unit of a commodity for units of another. Noting that MRS_{xy} represents the slope of the indifference curves for the two commodities X and Y, and that A and B are consumers, the Edgeworth box, in which the indifference maps of the two consumers are arrayed against one another, may be used to illustrate this proposition (Fig. 2.1). Suppose the initial endowment of commodities is at the point J with the consumers on indifference curves a_2 and b_2. It is clear that a Pareto improvement may be made by reallocating commodities in such a way as to move on to the 'contract curve' (AB), along which MRS_{xy} is the same for both consumers, between indifference curves a_2 and b_2. In this way one consumer will be better off and the other at least no worse off. At the point K, for example, both consumers are better off, being on indifference curves a_3 and b_3. Once a position is attained in which $MRS^a_{xy} = MRS^b_{xy}$ no further Pareto improvements are possible. Welfare cannot be further enhanced unless additional commodities are made available or judgements are introduced concerning the desirability of the distribution of commodities between the two consumers. For the moment we continue to leave distribution matters out of consideration.

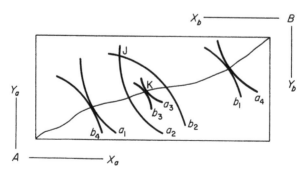

Figure 2.1   Efficiency in exchange.

Efficiency in production, the optimal allocation of inputs among producers, follows a similar mode of demonstration. This condition requires that the marginal rates of technical substitution (*MRTS*) between factors of production – the rate at which factors may be substituted for one another at the margin – be identical for all producers. Given such an outcome, no additional output can be secured from available resources. Noting that $MRTS_{kl}$ is the slope of the isoquants for two factors K and L and that X and Y are the producers (or commodities being produced), the Edgeworth box (here described in terms of isoquant rather than indifference maps) may again be employed for illustration (Fig. 2.2). If the initial allocation of factors is at the point J, output of one commodity may be increased without reducing the output of the other by reallocating factors between the production of X and Y until a position on the 'contract curve' is attained; for example, point K, between isoquants x_2 and y_2. Along the 'contract curve', $MRTS_{kl}$ is equal for both producers and Pareto optimality is achieved.

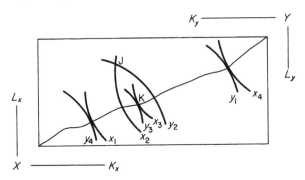

Figure 2.2 Efficiency in production.

Efficiency in both exchange and production, the optimal combination of outputs, requires that the MRS_{xy} for any consumer (and hence all consumers, given exchange efficiency in which MRS_{xy} is equal for all consumers) equals the marginal rate of transformation (MRT_{xy}) between the two commodities X and Y, where MRT_{xy} is the slope of the production possibilities or efficiency frontier derived from the 'contract curve' in Figure 2.2. Unless $MRS_{xy} = MRT_{xy}$, society would be willing to give up more than it is required to give up of one good for a unit of the other, and so would surrender some of the first good, enhancing social welfare in the process. Figure 2.3 illustrates this proposition.

At the point J, the representative consumer on indifference curve I would be willing to give up more of Y for an additional unit of X than the production possibilities frontier requires ($MRS_{xy} > MRT_{xy}$). Thus the consumer would give up Y until the point Z is reached, where the consumer attains the highest level of welfare consistent with production

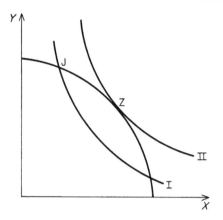

Figure 2.3   Efficiency in exchange and production.

possibilities. The point Z defines the optimal combination of commodities
to be produced.

2.2   PERFECT COMPETITION AND PARETO OPTIMALITY

It can now be demonstrated that in certain circumstances a perfectly
competitive market will automatically generate the three necessary
conditions of Pareto optimality as outlined in the previous section. The
fact that the circumstances in which this can occur are highly unrealistic
indicates the prevalence of 'market failure' and the consequent case for
government intervention in the economy and hence for the use of CBA.

The circumstances in which the perfect market yields Pareto optimality
are the absence of externalities (effects on third parties), the absence of
public goods and the absence of increasing returns to scale (see
Section 2.3 below). If we now make the standard assumptions of utility
maximization on the part of consumers and profit maximization on the
part of producers, given the absence of externalities, public goods and
increasing returns, Pareto optimality will automatically result from a
perfectly competitive market setting.

The proof is as follows. For utility maximization, each consumer sets
$MRS_{xy} = P_x/P_y$ and, since in perfect competition P_x/P_y is the same for all
consumers (no consumer being important enough in the market to
influence prices alone), it follows that MRS_{xy} is the same for all
consumers. This is the requirement for optimality in the allocation of
commodities among consumers (efficiency in exchange). For cost
minimization, each producer sets $MRTS_{kl} = P_k/P_l$ and, since in perfectly
competitive factor markets P_k/P_l is the same for all producers (no
producer being important enough to influence prices alone), it follows

that $MRTS_{kl}$ is also the same for all producers. This is the requirement of optimality in the allocation of inputs among producers (efficiency in production). Finally, for profit maximization, each producer in perfect competition sets $P = MC$ (marginal cost) so that $MC_x/MC_y = P_x/P_y$. Since $MRS_{xy} = P_x/P_y$ (as shown above regarding efficiency in exchange) and $MC_x/MC_y = MRT_{xy}$, it follows that $MRS_{xy} = MC_x/MC_y = MRT_{xy}$[1]. This is the requirement for the optimal combination of outputs (efficiency in both exchange and production).

Since we employ the notion in following sections of this chapter, it is emphasized that in the absence of externalities, public goods and increasing returns, the equality of price and marginal cost is a sufficient condition for the achievement of economic efficiency (or optimality). This may be illustrated by reference to the standard diagram of the perfectly competitive industry in Figure 2.4. Assuming away external effects, the demand curve, or price line, may be interpreted as the marginal social benefit curve (private and social benefits being coincident), given that it indicates the maximum amount that consumers (or beneficiaries) of the commodity are willing to pay for each unit purchased. The supply curve, or horizontal summation of individual firms' marginal cost curves, may be interpreted as the marginal social cost curve (private and social costs being coincident).

In perfect competition, the quantity Q_0 is produced and is sold at the price $P_0 = MC_0$, this quantity being socially optimal in so far as output is expanded up to the level at which marginal social benefit measured by price just equals marginal social cost. At $Q_1 < Q_0$ where $P_1 > MC_1$ a 'deadweight' loss of welfare is incurred of the order of the triangle abc, showing the amount which consumers would be willing to pay for units $Q_0 - Q_1$ over and above the costs of producing these units. This excess of benefit over cost on the units not produced represents a welfare loss to society. The level of output Q_1 is, therefore, inefficient. At $Q_2 > Q_0$

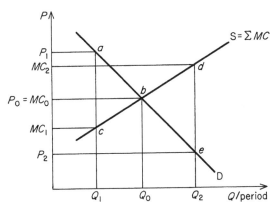

Figure 2.4 Marginal cost pricing.

where $P_2 < MC_2$, the cost of producing additional units beyond Q_0 exceeds by *dbe* the benefit of the additional units measured in terms of willingness to pay for them. Thus the level of output Q_2 is also inefficient. The condition $P = MC$, in other words, as generated in perfect competition under specified restrictive conditions, provides a rule for determination of the optimal level of provision of a commodity.

2.3   MARKET FAILURE

While perfect competition can thus be shown to generate Pareto optimality, there exist in practice situations in which even a competitive market system fails to provide the efficient level of output of a good or service. These are situations in which CBA may be considered as a replacement for the market so far as determination of the optimal scale of provision is concerned. They are ones involving externalities, public goods and increasing returns to scale. Furthermore, in conditions of imperfect competition, the market will not generate the optimal level of provision either. And, finally, optimality may be more completely defined in terms of income distribution as well as economic efficiency, in which event market signals often again fail. These are all situations suitable for the use of CBA.

2.3.1   Externalities

The first situation in which the competitive market fails is in the presence of external benefits or costs which impact upon parties other than the immediate consumers or suppliers of a commodity so that marginal private benefit does not equal marginal social benefit ($MPB \neq MSB$) or marginal private cost does not equal marginal social cost ($MPC \neq MSC$). Examples of external benefits include the benefits to others of a private vaccination or of education provided for certain groups in society. An example of external cost is air or noise pollution created through the process of private manufacture or travel. Figure 2.5 illustrates the extent to which non-optimal provision occurs because private revenues and expenditures fail to reflect social benefits and costs. Recognizing that the demand curve for a commodity represents the marginal private benefit curve (willingness-to-pay curve), the case of external benefit is illustrated in Figure 2.5a and of external cost in Figure 2.5b. In the first case, the optimal level of provision at which $MSB = MSC$ (Q_o) is greater than the level which would be provided through the private competitive market (Q_p) because private benefit is less than social benefit; in the second case $Q_o < Q_p$ because private cost is less than social cost. If the optimal level of provision is to be secured, it is therefore necessary to evaluate the extent of the externality in each case, a task which may be attempted by use of CBA.

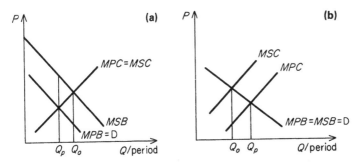

Figure 2.5 External benefits and costs.

2.3.2 Public goods

The market also fails to generate optimality in the presence of public goods. These are characterized by non-rivalry in consumption (A's consumption does not reduce B's) and by non-exclusion (B cannot be excluded from consuming the good even though it is provided for A). Examples include lighthouses and national defence systems. In the urban context, such things as street lighting, road maintenance, snow clearance, police and fire services display elements of publicness even if they are not in every instance 'pure' public goods in the sense of involving complete non-rivalry and non-exclusion.

Taking the extreme case of a 'pure' public good for purposes of illustration, the good affords non-depletable benefits for everyone from each unit produced. This means that the marginal valuation of an additional unit of the good is determined, not by what it is worth to the highest bidder (as in the case of an ordinary private good), but by its aggregate value to all consumers. In turn, this means that the market demand curve for a public good is derived by summing individual marginal willingness-to-pay (or demand) curves *vertically*, rather than *horizontally* as in the private good case. Thus, the optimal level of provision of the public good (that level of output at which marginal willingness-to-pay equals marginal cost) differs, except by coincidence, from the optimal level of provision that would prevail if the good were private.

These points are illustrated in Figure 2.6, using individual demand curves for consumers A and B. *DM* is the market demand curve. The optimal level of provision (Q_0) is shown by horizontal summation of individual demand curves for a private good in Figure 2.6a and by vertical summation for a public good in Figure 2.6b.

The problem in practice with public goods is that in order to arrive at the optimal level of provision each consumer's marginal evaluation has to be known. However, given non-rivalry and non-exclusion, consumers are

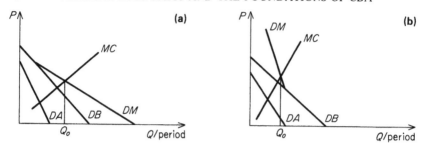

Figure 2.6 Private and public goods: optimal levels of provision.

unlikely to reveal preferences fully. Each consumer has an incentive to understate his true willingness-to-pay for the good (to take, if not a complete free ride, at least a cheap one) for he knows that, without paying for it, once the good is supplied he will not be excluded from consuming as much of it as he wishes. In the extreme event involving complete non-revelation of preferences by each consumer, the good will not be provided at all through the market system, and CBA may be used to determine whether it would be worthwhile, considering total economic benefits and costs, for the government to supply the good. In less extreme events, underprovision will occur. We might thus find an inadequate level of fire or police protection if these services are provided entirely through the private market. In such circumstances the government is again required to step in. In principle, CBA may then be employed to estimate the economic or social benefits and costs of different levels of service with a view to establishing a level closer to the optimum than would be achieved through the market.

2.3.3 Increasing returns to scale

There are two ways in which the existence of scale economies prevents the competitive market from generating the optimal level of provision. The first results directly from the natural monopoly case in which the market becomes monopolized through the incentive to exploit to the full the advantage of scale economies. With the loss of a competitive market, price will be set above marginal cost and quantity supplied will at the same time fall below the level at which $P = MC$. The second problem with increasing returns is that, even if somehow price were set equal to marginal cost (perhaps through public regulation), this would result in operating losses for the firm, and eventual closure, because $P = MC < ATC$ (average total cost). This is illustrated in Figure 2.7.

The falling average total cost curve over the relevant range of the demand curve indicates increasing returns, the result of high fixed costs relative to variable costs. The optimal level of provision is at Q_0 where $P_0 = MC < ATC$, creating an operating loss of P_0P_1ab. Pertinent

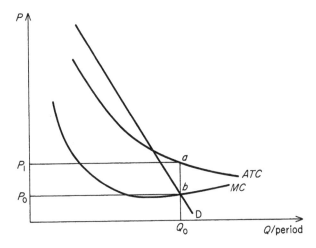

Figure 2.7 Increasing returns to scale.

examples might include construction and operation of a new municipal library or swimming pool.

If this undertaking is to proceed at the optimal level of provision, it will be necessary either to have the public sector assume responsibility for its establishment and operation, accepting the losses to be incurred, or to have the government subsidize a private concern to the extent of losses. In either case, CBA may be useful as an aid in determining whether the undertaking is economically worthwhile, considering total benefits and costs, despite commercial non-viability at the optimal price and quantity.

2.3.4 Other reasons for government intervention

There may be cause for the use of CBA on grounds other than the existence of externalities, public goods and increasing returns. For one thing, the market is unlikely to be perfectly competitive in the first place, even if increasing returns do not give rise to a natural monopoly situation. Thus, economic efficiency is not achieved, and a 'deadweight' loss of welfare exists to the extent that $P > MC$ (see Section 2.2). Market imperfections are addressed through policies of regulation, nationalization and anti-monopoly legislation; and CBA may in principle be employed to assess the net benefits of such intervention. In addition, governments may wish to intervene in the market, whether competitive or not, in order to influence the distribution of income and wealth as they do in the case of various social support programmes, including regional policies. In these cases, it is useful to establish to what extent efficiency is sacrificed in the interests of distributional equity. (For a discussion of the use of CBA in helping to quantify the efficiency impact of regional policies, see Chapter 15.) CBA may also be used to evaluate the relative efficiency of policy alternatives directed at influencing the distribution of income and wealth.

2.4   THEORY OF SECOND BEST

For all the foregoing justification for the use of CBA, the theory of second best (Lipsey & Lancaster 1956) presents a serious challenge to its worth. The general idea is that piecemeal attempts to force fulfilment of the optimality conditions may not enhance social welfare if any of these conditions remain unfulfilled elsewhere in the economy.

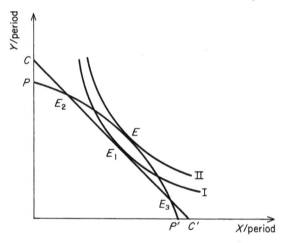

Figure 2.8   Second-best problem.

Figure 2.8 illustrates the point for a simple two-good world. A production possibilities or transformation curve (PP') is drawn for commodities X and Y with the point E indicating the 'optimum optimorum' at which $MRS_{xy} = MRT_{xy}$. Monopoly imperfections in the economy, however, create inefficiencies that prevent society from operating on PP' between E_2 and E_3 and thus prevent achievement of E. In effect, a constraint (CC') confines the feasible region of production to the frontier $PE_2E_1E_3P'$. This, then, is a world in which no more than a second-best outcome is possible. Now suppose that society finds itself at the point E_1 and that elimination or regulation of monopoly in sector Y could move society along CC' in the direction of E_2 on the production possibilities frontier. Similarly, a policy designed to attain Pareto conditions in sector X could move society in the direction of E_3. From the point of view of moving closer to the production possibilities frontier on which efficiency in production is achieved ($MRTS_{kl}$ is equal in both sectors), either move might seem to be desirable. Yet each move puts society on a *lower* social indifference curve than at E_1. It is better, therefore, to stay at E_1. In sum, when one of the marginal conditions for Pareto optimality is unattainable ($MRS_{xy} \neq MRT_{xy}$ in this example) it may be better to violate some of the other conditions (the equality of $MRTS_{kl}$ between producers in this example) since an apparent Pareto improvement can cause divergence from global optimality.

This theorem strikes at the very roots of CBA, designed as the method is to be an aid in the process of making recommendations for piecemeal improvements to social welfare. The argument generally advanced to defend implementation of piecemeal micro-economic initiatives in light of this problem is that the theory of second best depends on the existence of close interdependencies between all sectors in the economy (such that monopoly elsewhere, for example, really does affect the situation under analysis) and that, in practice, such close interdependence rarely prevails (e.g. Davis & Whinston 1965, Dasgupta & Pearce 1972, Ch. 4). For instance, imperfect competition in the car manufacturing industry is unlikely to be of practical significance when a proposal to improve the irrigation system in agriculture is under analysis.

2.5 SOCIAL DECISION RULES

If CBA is to be used as an aid to social decision making, it is necessary to have a framework in which it is possible to state whether a proposal is worthwhile or not and, among a set of proposals, how they are ranked. Proposed decision rules for this purpose are outlined in this section.

2.5.1 *Pareto criterion*

A first rule is the Pareto improvement criterion (see earlier) according to which a proposal will represent a social improvement if someone can be made better off without making someone else worse off. In this event, aggregate welfare will without question be enhanced. Unfortunately, however, this criterion is too restrictive for practical use, given that proposals generally involve gains to some people and losses to others. Where there are both gainers and losers the strict Pareto criterion becomes inapplicable.

2.5.2 *Kaldor–Hicks criterion*

As a way around this difficulty, the *potential* Pareto improvement criterion, or the Kaldor–Hicks (K–H) hypothetical compensation test (Kaldor 1939, Hicks 1939), has been suggested. This rule indicates that a proposal is acceptable if the gainers would be in a position to compensate the losers after implementation and still remain better off. This amounts to identifying as worthwhile a proposal for which aggregate benefits exceed aggregate costs, irrespective of *who* the gainers are or *who* the losers are. In cases of multiple proposals, schemes are ranked in terms of aggregate net benefit. This criterion constitutes the basis for traditional or 'economic' CBA (e.g. Krutilla & Eckstein 1958, Eckstein 1961b) which does not address the question of the distribution of benefits and costs, only their aggregate value.

Two difficulties are seen as attaching to the K–H criterion. The first is the so-called Scitovsky paradox (Scitovsky 1941). Scitovsky pointed out that while a first proposal could pass the K–H test, a second proposal to reverse the first could also pass the test. The reason is that in the process of implementing the first proposal, the structure of relative prices could be altered as a result of changes in relative supplies of goods and in factor incomes, the latter affecting the structure of demand for goods. This is illustrated in Figure 2.9. Society is initially located at Q_1 on price line I' (the commodities X and Y are shown on the axes). A move to the point Q_2 would be approved under the K–H test since Q_2 is clearly on a higher price line representing higher valued bundles of X and Y. Having moved to Q_2, however, a new set of relative prices comes into operation, given changes in commodity supplies and the structure of demand occasioned by the move. The ratio of the new prices is shown in the slope of price lines II' and II''. Society is now on curve II' and it is clear that a move back to Q_1 would be worthwhile, Q_1 being on a higher price line than Q_2 under the new price regime.

This paradox presents a dilemma for decision makers. In practice, however, the structure of relative prices may often be assumed to remain essentially unaffected by the sort of proposals analysed with CBA, so that the Scitovsky paradox may not represent a major problem.

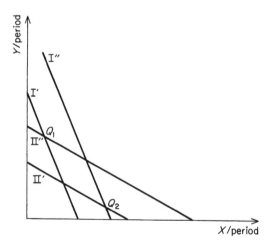

Figure 2.9 Scitovsky paradox.

The second difficulty with the K–H criterion is that it is strictly an efficiency criterion which offers no guidance to decision makers on the distributional effects of proposals. This is the case because benefits and costs are aggregated without regard to the question of who receives the benefit and who pays the cost. A dollar of benefit or cost counts equally, no matter to whom it accrues. Thus the rich may benefit and the poor lose, but that is not a consideration weighed explicitly under the K–H test. Moreover, there is no requirement that compensation actually be paid to losers in order to correct any undesirable distributional effects of a proposal. The criterion requires only that it be *hypothetically* possible for compensation to be paid while leaving gainers with a residual benefit.

Like the Scitovsky paradox, the narrow efficiency focus of the K–H criterion need not be a fatal weakness. Analysis of proposals in terms of their efficiency implications alone does not mean that distributional considerations are necessarily ignored in the decision-making process. What it does mean is that decision makers are left to weigh the importance of likely non-efficiency effects against the measured efficiency ones. Indeed, there is an influential body of traditionalist opinion which argues forcibly against attempts to go beyond use of the K–H criterion by incorporating non-efficiency considerations directly into the decision criterion (Musgrave 1969, Harberger 1971, Mishan 1974). The argument, explored in more detail in Chapter 6, is essentially that explicit inclusion of arbitrary distributional judgements in the decision criterion may be highly misleading, open to abuse and unnecessary.

2.5.3   Little criterion

While the perceived shortcomings of the K–H criterion would not, therefore, appear to preclude its usage (and, indeed, have not done so), Little (1951) nonetheless proposed an alternative criterion. He suggested that an undertaking should be considered worthwhile if it satisfies the K–H test, does not involve the Scitovsky paradox and provides a 'good redistribution of wealth'.

It may be noted that use of the Little criterion requires an explicit value judgement in order to determine what constitutes a 'good redistribution of wealth'. The problem then arises as to who should make that judgement and how. A further observation concerning the criterion is that value judgements are inescapable in the decision-making process for the reason that multiple objectives are involved and trade-offs have to be determined. It is not, therefore, a weakness of this criterion that it involves the need for an explicit judgement.

Used as the framework for application of CBA, this criterion clearly requires both economic and social analysis, the latter showing the impact on different groups in society. An undertaking is deemed acceptable on the basis of the K–H efficiency test, provided that Scitovsky reversability is absent and that a certain specified equity effect (a good restribution of wealth) is achieved. In matters of ranking, efficiency pay-off governs rank, while to be acceptable for inclusion in the ranking an undertaking has to satisfy both reversability and equity tests.

2.5.4 Bergson social welfare function

An alternative to the Little approach, and one which is comprehensive in the sense of encompassing explicit recognition of multiple objectives, is to employ a Bergson social welfare function (Bergson 1938) as the basis for application of CBA. This approach involves differential weighting of net benefits accruing to the different parties affected by a project or programme. Social welfare (SW) is specified as a function of net benefits (NB) accruing to each party, so that in linear form:

$$SW = \alpha_1 NB_1 + \alpha_2 NB_2 + \cdots + \alpha_n NB_n \qquad (2.1)$$

where the weights (α) reflect the contribution to different social objectives of creating a dollar of net benefit for each individual or group involved. Thus a dollar of net benefit for a privileged group is weighted less heavily than a dollar of net benefit for an underprivileged group if distributional equity is important as an objective, the degree of differential weighting reflecting the degree of importance of equity *vis-à-vis* efficiency. Any manner of objectives may be taken into account in this way. An undertaking is acceptable if the weighted sum of individual net benefits is positive, while the best schemes reveal the highest weighted net benefits.

While providing a broad and flexible framework for the implementation of social CBA, the Bergson social welfare function is not without difficulties in practice, for there remains the task of establishing values for the differential weights. We shall have more to say on this subject in Chapter 6. But it should be pointed out here that there are essentially two ways available for arriving at appropriate weights. The first is through the democratic process, either indirectly through elected representatives or directly by a process of majority voting. The practical problem of mobilizing voters to reveal goal preferences in the latter approach is one thing; the logical possibility of being unable to discern a clear ordering of goals even if people are effectively mobilized, or even if the matter is left to elected representatives, is another. This last problem is known as Arrow's Impossibility Theorem (Arrow 1963).[2]

An alternative approach to the determination of differential weights is to leave the task to specialists, either bureaucrats with the assumed necessary experience or the analyst alone. Unless the judgement of specialists is subject in some way to popular review, this obviously raises questions of appropriateness, given the canons of democratic decision making. It also raises questions in regard to the principles of positive economics according to which economists are not supposed to exercise personal judgement. Furthermore, the approach leaves unresolved the question of how specialists might go about their task of establishing weights, an issue dealt with in Chapter 6.

2.6 SUMMARY

This chapter has developed the foundations of CBA as they exist in the principles of welfare economics. It has been shown that the need for CBA arises as a result of government intervention in the provision of goods and services in circumstances where the private market fails to generate the socially optimal level of provision. In other words, CBA is used to guide resource allocation decisions where market signals fail. Circumstances in which failure of even a competitive market occurs include the presence of externalities, public goods, increasing returns and social goals other than maximization of efficiency in resource use.

Criteria available for guiding decision makers on the question of the absolute and relative desirability of projects and programmes include the Pareto, Kaldor–Hicks, Little and Bergson criteria. While the Pareto criterion lacks practical significance, the Kaldor–Hicks rule provides the framework for traditional economic CBA; and the Little criterion and the Bergson social welfare function provide frameworks for social CBA. Each of the criteria has limitations. Problems associated with second-best theory and Scitovsky's paradox are not generally considered to be of great importance in practice, provided that partial equilibrium analysis gives a sufficiently acceptable representation of the real world.

NOTES

1 That $MRT_{xy} = MC_x/MC_y$ is demonstrated as follows. The slope of the production possibilities curve (MRT_{xy}) equals the ratio of the amount of Y sacrificed to the amount of X gained when a small amount of any factor (say K) is transferred from the production of Y to the production of X. The amounts of Y and X respectively sacrificed and gained are measured by the marginal productivity of the factor (MP_k) in the production of each commodity. Thus:

$$MRT_{xy} = dY/dX = MP_{ky}/MP_{kx}$$

In turn, given that factor price (P_k) is market-determined for any producer in perfect competition:

$$MC_y \quad = (P_k \; dK)/dY = P_k/MP_{ky}$$

and

$$MC_x \quad = (P_k \; dK)/dX = P_k/MP_{kx}$$

So:

$$MRT_{xy} = (P_k/MC_y)/(P_k/MC_x) = MC_x/MC_y$$

2 Arrow's Impossibility Theorem demonstrates the possibility that a clear ordering of goals may not be generated by democratic majority preference voting, given the requirement of transitivity that if X is preferred to Y (XPY) and YPZ, then XPZ. Consider the case of three individuals (A, B, C) and three alternatives or goals (X, Y, Z). Ranking of goals in terms of individual preferences (from 3 down to 1) gives the following outcome:

	Goals		
Individuals	X	Y	Z
A	3	2	1
B	1	3	2
C	2	1	3

In this case, by majority XPY (individuals A and C) and YPZ (individuals A and B), so that by transitivity XPZ. By majority, however, ZPX (individuals B and C). It is clear that in a case such as this, where preferences are said to be non-single-peaked, a fundamental inconsistency obtains. If the alternatives in question were to be reduced to a single preferred option, by a procedure of pairwise elimination, the result would vary with the sequence of comparison.

3
Investment decision criteria

Before examining the details of benefit and cost measurement in the remaining chapters of Part I, we review the grounds for deciding on the acceptability and ranking of projects and programmes. The material in this chapter brings together the principles of investment appraisal as they apply to both public and private sector decision making.

The basis of the discussion is the so-called 'fundamental rule' (Stokey & Zeckhauser 1978): that in any situation involving project choice, the proposal (or group of proposals) to be selected is the one which produces the greatest net benefit. It should be clear that this rule derives directly from the Kaldor–Hicks test outlined in the previous chapter and is fully consistent with the use of a Bergson social welfare function involving differential weights. In the latter case, the 'fundamental rule' prescribes maximization of differentially weighted net benefits. If the Little criterion is employed for public sector decision making, then the rule is modified, but not basically changed, by operation of the equity constraint. In this chapter we discuss the application of the 'fundamental rule' through use of subsidiary project appraisal criteria. But, first, we outline the techniques of compounding to terminal value and discounting to present value as these concepts are important in investment decision making.

3.1 TERMINAL AND PRESENT VALUES

Providing that interest is reinvested annually, a sum of money (X_0) grows at compound interest (i) to an increasing amount in future years. The amount to which it grows (X_t) at the end of each subsequent year is given as $X_0(1+i)^t$ where t represents the number of years over which the sum X_0 is compounded. The value at the end of whatever period is deemed appropriate is the terminal value of the original sum. Conversely, the present value of a future sum is that amount (X_0), paid or received today, which is the equivalent of the future sum. Given that $X_t = X_0(1+i)^t$, it follows that $X_0 = X_t(1+i)^{-t}$.

For purposes of project analysis, we can, therefore, define the terminal value of a stream of annual benefits or costs over a project's life (t) as:

$$X_0(1 + i)^t + X_1(1 + i)^{t-1} + X_2(1 + i)^{t-2} + \ldots + X_t(1 + i)^0 \quad (3.1)$$

and the present value of a stream of annual benefits or costs as:

$$\frac{X_0}{(1 + i)^0} + \frac{X_1}{(1 + i)^1} + \frac{X_2}{(1 + i)^2} + \ldots + \frac{X_t}{(1 + i)^t} \qquad (3.2)$$

where the terms $1/(1 + i)^0 \ldots 1/(1 + i)^t$ are discount factors, widely available in computed tabular form for easy reference.

If the annual flow of benefits or costs is constant (X), then the terminal value of the stream may be expressed as:

$$\frac{X\,[(1 + i)^t - 1]}{i} \qquad (3.3)$$

and the present value as:

$$\frac{X\,[1 - (1 + i)^{-t}]}{i} \qquad (3.4)$$

If the stream also occurs in perpetuity (or it is convenient to assume an infinite time horizon in the case of very long-term projects), terminal value is no longer relevant, but present value becomes X/i.

The principles of compounding and discounting highlight the significance of the time value of money: money received or paid early is worth more than money received or paid late because it can be put to work to earn interest. This crucial fact is to be recognized in devising suitable criteria for appraising investments which generate flows of costs and benefits over time. We now consider available criteria and their use in different decision situations.

3.2   LEADING INVESTMENT CRITERIA

There are three criteria widely employed in investment decision making: the net present value (NPV) criterion, the benefit–cost ratio $(B/C$ ratio) and the internal rate of return (IRR).

NPV is measured as the present value of benefits (PVB) less the present value of costs (PVC), where benefit and cost streams are discounted at the minimum return requirement (MRR), or opportunity cost rate of return on the resources employed in the project. Details concerning the establishment of the MRR are discussed in Chapter 8. The B/C ratio is measured as PVB/PVC, again discounting at the MRR. The internal rate of return is that rate of discount applied to benefit and cost streams which sets $PVB = PVC$, or $NPV = 0$. The IRR may alternatively be seen as the rate of return generated on the outstanding capital in each year of the life of a project. A simple example will illustrate.

Table 3.1 Project cash flows.

Year	Loan outstanding (opening balance) $	Interest 8% $	Debt year end $	Repayment from net cash flow $	Loan outstanding (closing balance) $
1	1000	80	1080	388	692
2	692	55	747	388	359
3	359	29	388	388	–

Suppose we borrow $1000 at 8% in order to undertake a project expected to provide a net cash (benefit) flow of $388 p.a. for three years. At the end of each year we reduce the amount of the loan outstanding by the value of the net cash flow. Table 3.1 presents the schedule of receipts and payments. In words, the project yields just enough to repay the loan (recover the capital cost) and to provide interest of 8% on the outstanding balance of the loan (or the capital) each year. The project's IRR is 8%. Looking at it in terms of setting $NPV = 0$, it is seen that $PVB = PVC$ when inflows and outflows are discounted at 8%:

$$PVC = \$1000$$

$$PVB = \frac{\$388}{(1.08)} + \frac{\$388}{(1.08)^2} + \frac{\$388}{(1.08)^3} = \$359 + \$333 + \$308 = \$1000$$

Thus, the IRR may be determined as the rate of discount (i) at which $NPV = 0$.

In the following two sections we examine the use of the three leading criteria in two different situations: one in which the acceptability of a single undertaking is to be determined, and one in which multiple proposals are to be ranked in order of desirability.

3.3 SINGLE ACCEPT/REJECT DECISION

In this situation a single project or programme is analysed to determine whether or not it is worthwhile. Acceptability requires that $NPV > 0$ or $B/C > 1$, discounting at the MRR. In many circumstances, too, the proposal is worthwhile if $IRR > MRR$. However, it is to be noted that the IRR loses uniqueness and more than one rate emerges when there is more than one sign change in the net benefit stream. This is because the

IRR is the solution to a polynomial equation which will have a root for every degree of the equation. For example, the net benefit flow over three years of −$1000, +$2550, −$1575 beginning in year 1 has two *IRR*s at 5% and 50%. The present value profile of this dollar stream showing *NPV* at different discount rates (*i*) is displayed in Figure 3.1. It can be seen that use of the *IRR* criterion indicates acceptability if *MRR* < 5%, even though *NPV* < 0 at discount rates below 5%, clearly a violation of the 'fundamental rule'.

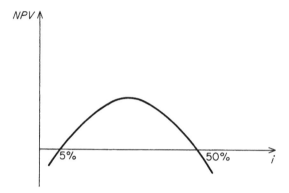

Figure 3.1   Present value profile: double IRR.

A characteristic of the three criteria under consideration is that each involves what may be an unrealistic implicit assumption concerning the reinvestment of project proceeds during the life of a project. Specifically, the *NPV* and *B/C* rules assume that proceeds are reinvested at the *MRR* used as the discount rate, while the *IRR* assumes reinvestment at the internal rate of return. Clearly, once the possibility of consuming annual benefits is recognized, as for example occurs in the case of non-cash recreational or time-saving benefits, then any reinvestment assumption may be inappropriate. Moreover, even in the absence of benefit consumption, there may be no reason to suppose that reinvestment opportunities are available at the internal rate of return of the project in question even if reinvestment at the *MRR* (the opportunity cost of capital) remains a reasonable assumption.

The presence of the reinvestment assumption is easily demonstrated. For *NPV* and *B/C*, a project is worthwhile if:

$$\sum_{t=0}^{n} \frac{B_t}{(1+i)^t} > \sum_{t=0}^{n} \frac{C_t}{(1+i)^t} \tag{3.5}$$

Multiplying through by $(1+i)^n$ gives

$$\sum_{t=0}^{n} B_t(1 + i)^{n-t} > \sum_{t=0}^{n} C_t(1 + i)^{n-t} \qquad (3.6)$$

indicating that if the stream of benefits compounded forward to period n at the rate i is greater than the stream of costs similarly compounded, the project is worthwhile. Expression (3.6) is formally equivalent to expression (3.5) and makes explicit the assumption of reinvestment (or compounding) at the rate i. The same demonstration applies to the *IRR*, where i is that discount rate at which:

$$\sum_{t=0}^{n} \frac{B_t}{(1 + i)^t} = \sum_{t=0}^{n} \frac{C_t}{(1 + i)^t} \qquad (3.7)$$

In order to circumvent the reinvestment problem it is necessary to have information regarding the proportion of annual benefits reinvested. With this information, the most convenient procedure is to compound benefits forward to the terminal year of project life using the minimum return required on investment in respect of the proportion reinvested each year, and the minimum return required as compensation for forgone consumption in respect of the proportion consumed (see Mishan 1982a, Chs 27, 28).[1] Costs are also compounded forward so that benefits and costs may be compared at terminal rather than at present value. This procedure has not yet been widely adopted, the information requirement regarding the reinvestment proportion and the separate interest rates being somewhat severe. It should be recognized, however, that continued use of the conventional leading criteria may typically involve a measure of approximation.

3.4 RANKING OF PROJECTS

When it is necessary to rank projects in order of preference, either because they are mutually exclusive or because a budget or other constraint precludes adoption of all proposals which pass the accept/reject test, the *NPV* criterion may give conflicting results with the *IRR* and *B/C* ratio, unless these latter criteria are used properly. The conflicts are illustrated in Figure 3.2. In Figure 3.2a the present value profiles of projects A and B are depicted. In terms of *IRR*, B appears to be the preferred project. However, in terms of *NPV* this is the case only in the event that $i = MRR > i_x$. At any $MRR < i_x$, A is superior. In Figure 3.2b, project A, the larger project, gives a greater *NPV* but a lower *B/C* ratio.

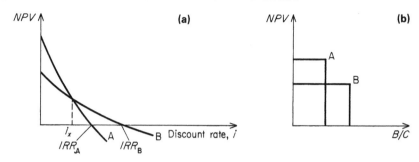

Figure 3.2 Decision criteria conflicts.

In the following subsections we discuss the resolution of these conflicts in the separate cases of mutual exclusivity and budget or other constraints.

3.4.1 Mutual exclusivity

In the case of mutually exclusive projects, *NPV* clearly provides a solution consistent with the 'fundamental rule'. The problem with the *B/C* ratio and *IRR* is that neither gives consideration to the absolute value of net gain, and thus the possibility of conflict with *NPV* arises. In order to use these latter criteria in the mutually exclusive decision situation, it is necessary to conduct an incremental analysis, as illustrated below.

Table 3.2 shows that according to the *NPV* criterion project B is preferable, discounting at *MRR* = 8%. However, the *B/C* ratio and *IRR* criteria suggest that project A is preferable. These latter criteria can be seen to be misleading if we ask whether it is worthwhile (a simple accept–reject decision) to invest the additional capital cost of $278 in project B. The third row of the table indicates, by all three criteria, that the incremental investment (call it project C) is worthwhile. The table shows that hypothetical project C carries a positive net benefit of $17 and is acceptable according to *B/C* and *IRR* criteria. This means that it is worthwhile to invest the additional $278 in project B. In other words, B is preferable to A. It is only through such incremental analysis that the *B/C* ratio and *IRR* provide the correct guidance, subject to the caveat concerning the reinvestment assumption.

Table 3.2 Comparison of two projects

Project	Capital cost in Year 0 $	Annual net benefit flow $	Project life (years)	NPV at 8% $	B/C at 8%	IRR(%)
A	502	100	10	169	1.34	15.0
B	780	144	10	186	1.24	13.0
C	278	44	10	17	1.06	9.4

Thus, if we wish to employ the *B/C* and *IRR* criteria in respect of mutually exclusive decisions it is necessary to analyse the incremental return or ratio between projects.[2] This is obviously a more cumbersome procedure than is involved in use of the *NPV* criterion and one which may become very complicated in the presence of multiple alternatives where a series of sequential pairwise comparisons is required.

In the more complicated case of projects which are interdependent so that the pay-off from one depends on implementation of another, the *NPV* rule still applies. However, it is necessary to analyse all possible combinations of projects to identify the one with the greatest *NPV*.

Finally, it should be noted that two situations which are common in practice are special cases of mutual exclusivity. These involve questions of optimal scale for a proposed development and optimal timing. In both cases, alternatives are analysed as separate proposals. In the second case, it is important to recognize that net benefits may be increased as a result of delay either because the present value saving in capital costs may outweigh the present value reduction in annual net benefits, or else postponement may provide improved information concerning estimated costs and benefits.

3.4.2 Capital budget or other single period constraint

When capital funds are limited to a budgeted amount, it is important to rank proposals in order of preference for purposes of determining which proposals to undertake within the capital constraint. The same consideration applies in face of any single period constraint, for example regarding the availability of resources required in the construction of a project. The 'fundamental rule' indicates that net gain is to be maximized over the group of proposals undertaken. This means that we seek to maximize the benefit per dollar expended on each proposal undertaken. Because relative 'profitability' is, therefore, relevant when choosing projects within the constraint, the most convenient criterion is the *B/C* ratio, as illustrated in Table 3.3 for the case of a capital budget constraint. Table 3.3 indicates that proposals are ranked A, B, C, D according to the *B/C* ratio. *NPV*, by contrast, ranks them D, B and C, A. The assumed constraint is $100 000.

Table 3.3 Alternative proposals

| | Proposals ($000) | | | |
	A	B	C	D
PV capital cost	10	40	50	100
PV annual net benefits	20	60	70	130
B/C	2	1.5	1.4	1.3
NPV	10	20	20	30

If the *NPV* criterion were adopted for ranking purposes, only proposal D could be undertaken within the constraint for a net gain of $30 000. Use of the *B/C* ratio, on the other hand, involves adoption of proposals A, B and C for an overall net gain of $50 000. The recommended procedure, then, is to rank by the *B/C* ratio and work down the list of proposals until the capital budget is fully used. Use of *NPV* involves the inconvenience of having to compare every combination of proposals which can be financed under the constraint.

An important qualification to the above argument, however, is that the simple rule of ranking by the *B/C* ratio becomes inoperative when the budget is violated as we move down the ranked list from one proposal to the next. In this situation, it is necessary to determine which proposal(s) to skip over and which to adopt, taking into consideration every combination of proposals that can be financed within the constraint. If the 'fundamental rule' is to be observed, there is no escape from the need to compare all feasible combinations in terms of *NPV*.

3.4.3 *Multiple constraints and other complications*

When there are multiple constraints in the initial period and/or financial or other resources are rationed over different periods, the problem may require the use of mathematical programming in order to maximize net present value subject to whatever complex constraints apply (Weingartner 1963). Integer programming may also be required in cases of indivisibilities where the 'lumpiness' of proposals precludes simple application of maximization rules based on continuous relationships (Weingartner 1963).

3.5 OTHER INVESTMENT CRITERIA

Decision criteria other than *NPV*, *B/C* ratio and *IRR* are sometimes employed. We discuss them in two categories: those which are consistent with the 'fundamental rule' and those which are not.

3.5.1 *Criteria consistent with 'fundamental rule'*

Two criteria fall into this category, both of which are formally equivalent to the *NPV* criterion. The first is the equivalent annual value (*EAV*) method. This involves conversion of a fluctuating annual flow of benefits, costs or net benefits into an equivalent constant annual flow, an annuity (*A*), such that for a net benefit flow for example:

$$\sum_t \left[\frac{A}{(1 + i)^t} \right] = \sum_t \left[\frac{NB_t}{(1 + i)^t} \right] \tag{3.8}$$

This expression indicates that the value of A is found by dividing the present value of ΣNB_t by the cumulative present value factor for $i = MRR$.

Table 3.4 Present value of net benefit stream

End year	$NB(\$)$	Discount factor (%)	$PV(\$)$
1	120	0.909	109.0
2	120	0.826	99.1
3	54	0.751	40.5
Total	294	2.486	248.6

Suppose we have the stream of net benefits as shown in Table 3.4 and $i = MRR = 10\%$. Then $EAV = \$248.6/2.486 = \$100$ p.a. for three years. That \$100 p.a. for three years is the equivalent of a flow of \$120, \$120 and \$54 may be checked as shown in Table 3.5.

Table 3.5 Present value of three-year annuity of \$100

End year	$A(\$)$	Discount factor (%)	$PV(\$)$
1	100	0.909	90.9
2	100	0.826	82.6
3	100	0.751	75.1
Total	300	2.486	248.6

The EAV can be useful for comparing alternatives involving annual non-monetary effects which require to be weighed against effects measured in monetary terms. An example of its use in this way is found in the Planning Balance Sheet method used for comparing alternative urban development plans (see Chs 6 and 12).

The second criterion in the category of additional methods which are consistent with the 'fundamental rule' is the net terminal value (NTV) approach referred to in Section 3.3 (Mishan 1982a). This involves compounding forward to the end of a proposal's life rather than discounting back to present value. Thus proposals are assessed in terms of net terminal rather than present value. We have explained in Section 3.3 that this criterion is in principle convenient for taking account of the reinvestment issue. Terminal values, of course, may also be used as the basis for computing the B/C ratio.

3.5.2 Criteria inconsistent with 'fundamental rule'

Other criteria sometimes employed in analyses are to be classified as inconsistent with the 'fundamental rule'. One example is the pay-back criterion by which proposals are ranked in terms of cost-recovery periods, a procedure which may well conflict with the objective of maximizing net benefit. A second example is the average rate of return, sometimes referred to as the accountant's rate of return (Merrett & Sykes 1966, p. 101), which ranks proposals in terms of average undiscounted annual net benefits (NB) as a proportion of capital outlay (K):

$$\left(\frac{1}{n} \sum_{t=0}^{n} NB_t \right) \Big/ K \qquad (3.9)$$

It is clear that this criterion ignores the importance of the time value of money, account of which is implicitly included in implementation of the 'fundamental rule'.

3.6 SUMMARY

Acceptable investment criteria stem from the 'fundamental rule', that decisions regarding proposals be made in light of maximizing net benefits (having regard for the time value of money). The appropriateness of different criteria depends on the type of decision to be made. In the case of a simple accept-or-reject decision in relation to a single project, the NPV, B/C, IRR, NTV or EAV rules are all satisfactory, subject to the qualifications that their implicit reinvestment assumptions may not be realistic, and that the IRR may not always be unique. The NTV rule (or B/C ratio, defined on the basis of terminal values) offers a convenient vehicle for dealing with the reinvestment issue, although information requirements remain severe.

Where projects need to be ranked in order of priority, the IRR and B/C rules are to be avoided in cases of mutual exclusivity, unless an 'incremental' analysis is conducted. In conditions of a single period constraint, the B/C rule may be used as the basis for ranking, unless project indivisibilities exist requiring that all feasible combinations of projects be compared in terms of NPV. In other, more complicated decision situations, mathematical programming methods may be required. Additional decision formulae which are inconsistent with the 'fundamental rule' are best ignored.

NOTES

1 The minimum return required as compensation is used to compound consumption benefits since society is indifferent between a return of this rate and current consumption; current consumption is therefore the equivalent of a future return of this order.
2 It is sometimes argued that decision makers prefer to have results expressed in rate of return terms (Merrett & Sykes 1966, Ch. 11).

4

Measurement of benefits and costs: welfare surpluses

In this chapter we begin an examination of the principles underlying measurement of benefits and costs in economic and social analysis. A money measure is required of the gains and losses in welfare associated with initiatives under investigation. If money incomes are changed directly by a policy initiative, the change in welfare is measured as the change in income. But if product prices are changed, if prices do not properly reflect the true economic value of the effects produced by the initiative, or if market prices are non-existent, the problem of measurement becomes more complicated.

The focus in this chapter is on the measurement of welfare effects when prices change. The next chapter deals with measurement when prices do not reflect true economic value and when market prices do not even exist.

4.1 MEASUREMENT OF BENEFITS

When product price is altered as a result of a project, programme or policy, consumers of the product experience a welfare change. If a policy decision reduces bus fares at off-peak periods of the day, for example, a certain group of bus travellers enjoys a welfare gain. The same change in price may also benefit or hurt the owners of factors of production (labour, land or capital) involved in supplying the product. In the case of the bus service, bus company profit (the return to capital) may be reduced by the change in fares. Thus, abstracting from external effects on third parties, the benefits of an initiative that causes a reduction in product price are the combined changes in welfare for both consumers (or users) and factor owners (including government if the product is publicly supplied).

For gainers, benefits are measured in terms of either their maximum willingness-to-pay for the enhanced welfare of the beneficial price change, or of the minimum compensation required to induce them to forgo the change. For losers, the loss of welfare is measured as either the minimum amount they are willing to accept as compensation for the loss or the maximum amount they are willing to pay to avoid it. These principles apply to consumers and factor owners alike. It might also be said that they extend as well to the measurement of external benefits for third parties.

To the extent that differential weighting is not applied to gains and losses, measures of willingness-to-pay and required compensation involve implicit assumptions that individual preferences count and that the prevailing distribution of income and wealth is acceptable. The first assumption allows benefits to be determined in terms of what individuals, rather than some omniscient body of government, consider them to be worth. The second assumption allows that ability-to-pay will influence willingness-to-pay, since the latter is in part dependent upon the former. The introduction of differential weights when these assumptions are deemed unacceptable is discussed in Chapter 6.

From the practical standpoint, welfare impacts on consumers and factor owners are measured respectively as changes in consumer and producer surpluses. Consumer surplus is the amount which consumers are willing to pay for a product over the price that they are required to pay. Alternatively, consumer surplus may be seen as the compensation required above product price in the event that the consumer is denied the product. Since the demand curve for a good or service indicates the price which consumers are willing to pay (or the compensation required in the event of denial) for each incremental unit of the commodity, consumer surplus is measured as the area under the demand curve and above the price line. Similarly, since the supply curve of a product or factor of production indicates willingness-to-supply (or minimum compensation required for supplying the product or factor service), producer surplus is measured as the area above the supply curve and below the price line.[1]

4.1.1 Consumer surplus measure of benefit

Beginning with the individual consumer in reference to Figure 4.1, total willingness-to-pay for Q_1 units (or compensation required if he/she is denied these units) is the amount $(a + b + d)$. At price P_1 expenditure for these units is represented by the area $(b + d)$ so that consumer surplus is (a). If price falls to P_2, total willingness-to-pay is $(a + b + c + d + e)$, expenditure is $(d + e)$ and consumer surplus is $(a + b + c)$. The benefit of the price fall measured by the change in consumer surplus is, therefore, $(b + c)$. Aggregating individual demand curves, either horizontally for a private good or vertically for a public good (see Ch. 2), generates the market demand curve from which the gain to all consumers may be measured in like fashion, reinterpreting Figure 4.1 as the market demand curve.

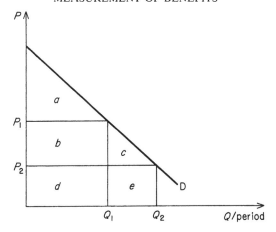

Figure 4.1 Product demand curve.

4.1.2 Producer surplus measure of benefit

In the context of an individual supplier, or of the industry as a whole, the impact of a price change on producer surplus is the area above the product supply (or marginal cost) curve and between product price lines. In Figure 4.2, producer surplus gain is shown as the area (a) for the case of a price increase in a competitive market (where the supply curve is the marginal cost curve above average variable cost). If the position of the MC curve is not affected by the price change of the product (that is, if variable factor prices are not changed), this area (a) measures the change in rent accruing to fixed factors, typically as residual income to capital (the shareholders in a private firm or the government in the case of a publicly supplied product).

In the case where variable factor prices do change along with product price, it is emphasized that the area (a) represents the net change in

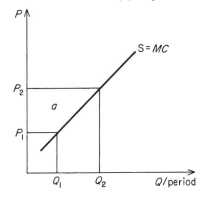

Figure 4.2 Product supply curve (competitive market).

producer surplus for all factors affected by the project, programme or policy (Sugden & Williams 1978, p. 257). In the event of an increase in product price, the demand for variable factors is likely to increase (the demand curve for a factor being the value of its marginal product curve when its output is sold in a perfectly competitive product market, or the marginal revenue product curve for an imperfect product market), creating an increase in producer surplus for these factors as their prices rise. This process is illustrated in Figure 4.3, using labour as the variable factor in Figure 4.3b. Perfectly competitive product and factor markets are assumed for illustration. Owners of fixed factors secure a gain of the area $(a + b)$ in Figure 4.3a less the area $(c + d)$ in Figure 4.3b, the latter amount being a gain of surplus for labour. Thus the area $(c + d)$ is self-cancelling as a transfer from capital or land to labour, the net change in producer surplus being the area $(a + b)$ in Figure 4.3a which corresponds with area (a) in Figure 4.2.

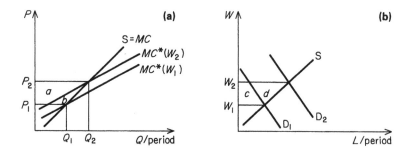

Figure 4.3 Producer surplus: net increase.

In practice, of course, it may be necessary to trace producer surplus to its accrual in separate factor markets (Pearce & Nash 1981, p. 99). For example, an industry supply curve is not defined in imperfect competition; or in the case of a social analysis, benefits and costs to different factors of production may need to be identified separately.

4.1.3 Combined surplus measures: total benefit

To the extent that producer surplus, however, can be identified from the product supply curve, it is useful pedagogically to demonstrate the measurement of aggregate economic benefit using supply (marginal cost) and demand curves. Taking by way of illustration a project which reduces product price and generates additional output, there are three cases of interest (Freeman 1975).

The first case involves conventional demand and supply (marginal cost) curves (Fig. 4.4). The scheme under analysis shifts the supply curve to the

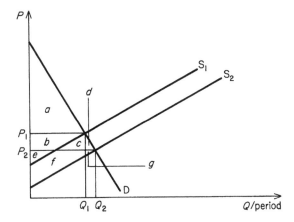

Figure 4.4 Project benefits: standard case.

right and reduces product price from P_1 to P_2. Examples could include the impact on an industry of introducing new machinery or improved methods of operation. The increase in consumer surplus/period = $(b + c + d)$. Producer surplus alters from area $(b + e)$ to area $(e + f + g)$, giving a change in producer surplus of $(f + g - b)$. Total benefit/period is, therefore, $(c + d + f + g)$. The area (b) cancels out as a transfer from producer to consumer surplus. Only if it were important (in a social analysis) to identify the distribution of gains and losses between different parties would it be important to measure (b).

The second case is the constant cost case, which is fairly common at the project level. One widely known version of it involves a road improvement scheme (see Ch. 10). In Figure 4.5 the 'generalized' cost/journey along the improved section of road is plotted on the vertical axis and includes vehicle operating, time and other costs of travelling. The number of journeys/period is plotted on the base axis. The road improvement provides a reduction in 'generalized' journey cost, or price. Users of the road who would have travelled along it anyway in the absence of the improvement secure a welfare gain measured as the area (a). In addition, new traffic is generated $(Q_2 - Q_1)$ as a result of the travel cost reduction, the benefits accruing to new travellers being the area (b), the excess amount they are willing to pay for the $Q_2 - Q_1$ journeys made $(b + c)$ over the cost of making these journeys (c). The addition to consumer surplus/period is, therefore, $(a + b)$. In this case the change in producer surplus is zero.

It may be noted that the change in producer surplus is not necessarily zero in the example of constant cost. Such is the case when excess profit is earned. The decision to reduce the off-peak bus fare referred to earlier may be used as an illustration. Suppose that the marginal cost of carrying passengers is constant and that before the fare reduction the bus company

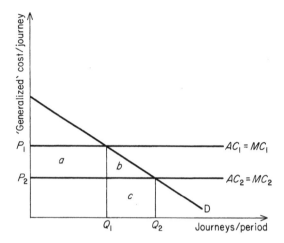

Figure 4.5 Benefits: road improvement scheme.

was earning excess profit from off-peak operations of the amount $(a + c)$ in Figure 4.6. The decision to reduce the fare from P_1 to P_2 creates additional consumer surplus of $(a + b)$ and a change in producer surplus of $(d - a)$. Total benefit/period is, therefore, $(a + b + d - a) = (b + d)$, the area (a) being merely a transfer from producer to consumer surplus.

The third case concerning benefit measurement when output is increased by a project is one where a single producer in perfect competition experiences a fall in marginal cost and an increase in quantity demanded. Here the change in consumer surplus is zero and net benefit is measured solely in terms of producer surplus $(a + b)$ (Fig. 4.7). In this case benefit accrues entirely as rent to factors of production. The real-world analogue of this case is when cost savings generated by a project are not passed on to consumers.

Three additional observations are in order concerning welfare surplus measures of benefit. The first is that, even though no actual price change occurs, the benefit to society of the very existence of a facility (say a public park) may be measured using the principles outlined. The benefit of a project which creates a new good or service can also be measured, again despite the fact that no price change is involved. The reason is that provision of the good or service involves a notional reduction in price from the level at which zero use occurs to the observed level. Benefit/period is, therefore, represented by the sum of consumer and producer surpluses created by the existence of the facility or product; that is, the whole area under the demand curve and above the marginal cost (or supply) curve.

The second additional observation is that, although the preceding discussion has related to a price change for a final product, the principles of benefit measurement outlined can also be applied to a price change for

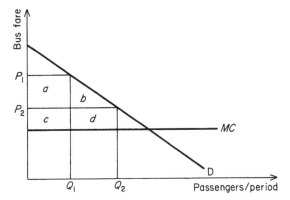

Figure 4.6   Benefits: reduced bus fares.

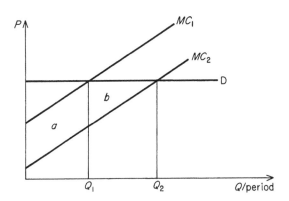

Figure 4.7   Benefits: single producer in perfect competition.

an intermediate product or production input. Construction of a new dam, for example, could reduce the cost of farmland irrigation. The benefits of the dam could be estimated from either demand and supply (marginal cost) functions in the final product markets for farm output, or from the demand curve for water on the part of farmers.

Finally, price changes used for illustration in this section have occurred as a result of supply shifts or, in the case of the bus fares example, policy edict. It is, of course, the case that the principles of benefit measurement also apply in connection with price changes resulting from demand shifts. An example is discussed in Chapter 9 where urban renewal schemes are seen as increasing demand for the fixed supply of land at a renewal site, creating thereby a benefit in terms of producer surplus for landowners.

4.2   MEASUREMENT OF COSTS

Costs are measured in terms of either the minimum amount sufferers are willing to accept as compensation for the loss of welfare created by the project, programme or policy, or the maximum amount they are willing to pay to avoid that loss. Obviously, reductions in consumer or producer surpluses (negative benefits) qualify as costs. In Section 4.1, reductions in producer surplus were netted out in the computation of benefits, a procedure which could also be followed, for instance, in respect of negative benefits to 'consumers' of externalities that result from a project. In the case of the road improvement scheme, for example, increased traffic may inconvenience neighbourhood residents, the cost to them being measured in terms of lost consumer surplus on 'peace and quiet'.[2]

Aside from these costs of lost consumer or producer surpluses, however, it is necessary to take account of the cost of resources absorbed by the project or programme under investigation. This cost comprises the opportunity cost of using resources for project construction and operating purposes, that is, the value which these resources would have produced in alternative use. Opportunity cost measures the welfare displaced by the project: the value of alternative output forgone as a result of it. This basis of evaluating the capital and continuing costs of a project requires in principle that forgone surplus on resource use in alternative employments be taken into account (Dunn 1967). In practice, however, we usually settle for forgone market value, subject to adjustments where appropriate (see Ch. 5).

4.3   CONSUMER SURPLUS: SOME DETAILS

Thus far we have not addressed the question of the precise delineation of the demand curve, the area beneath which and above the price line we recognize as consumer surplus. In this section we turn to that question. Following Hicks (1943–4), the analysis is developed in terms of evaluating the benefit or loss to the single consumer of a price change.

4.3.1   The four consumer surpluses

Figure 4.8 illustrates the example of a price fall. The quantity of good X, the price of which falls, is shown on the abscissa with the amount of income or money available for all other commodities on the vertical axis. The consumer's budget constraint rotates from HL to HM following the fall in the price of X, and utility maximizing equilibrium changes from A to B. Four separate measures of the change in welfare occasioned by this price fall may now be identified:

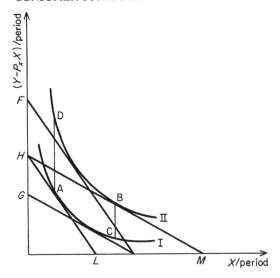

Figure 4.8 Hicksian consumer surpluses: single consumer.

(a) Compensating variation $(CV_1) = HG$: the maximum amount of money the consumer would be willing to pay in order to secure the price fall. If the amount HG were paid, the consumer would be no worse off than before the price fall, being on welfare level I.

(b) Compensating surplus $(CV_2) = BC$: the maximum amount of money the consumer would be willing to pay in order to secure the price fall, assuming that he/she is constrained to buy the quantity of X which would be bought were the price to fall and none of the real income gain extracted. Again the consumer would be no worse off than before the price fall, being on welfare level I.

(c) Equivalent variation $(EV_1) = HF$: the minimum amount of money the consumer would require in order to forgo the benefit of the price fall. Compensation of this order would place the consumer on welfare level II which he/she would attain were price to fall.

(d) Equivalent surplus $(EV_2) = DA$: the minimum amount of money the consumer would require in order to forgo the price fall, assuming that he/she is constrained to buy the quantity of X which would be bought in the absence of the price fall. Again the consumer would attain welfare level II.

Since CV_1 and CV_2 are measured in terms of the new price ratio while EV_1 and EV_2 are measured in terms of the original price ratio, it should be plain that in the case of a price increase involving loss of welfare the following measures are identified:

(a) $CV_1 = HF$: the minimum amount required as compensation for the price rise.

(b) $CV_2 = DA$: the minimum amount required as compensation for the price rise, assuming the quantity constraint.

(c) $EV_1 = HG$: the maximum amount the consumer would be willing to pay to avoid the price rise.

(d) $EV_2 = BC$: the maximum amount the consumer would be willing to pay to avoid the price rise, assuming the quantity constraint.

Silberberg (1972) has taken this analysis further, showing that an infinite number of welfare change measures exist, two for each quantity which the consumer could be constrained to purchase.

It should be noted that unless the indifference curves are parallel so that the income effect is zero, the separate measures of change in the welfare of the consumer are not equal. For a price fall, for example, $EV_1 > CV_1$ for a normal good, but $EV_1 < CV_1$ for an inferior good. Thus there is no unique measure of welfare gain or loss. The question then arises as to which measure should be employed in defining consumer surplus.

The answer depends on whether the consumer is considered to have no right to make purchases at the new price (in which case the CV is relevant) or to have that right yet to be denied the opportunity to buy (in which case the EV is appropriate). The answer, therefore, depends in part on a judgement as to which underlying distribution of property rights is the more equitable (Krutilla 1967, Mishan 1976). It also depends on whether or not the consumer is quantity-constrained, as, for example, he may be in the market for public housing (see Ch. 9). Thus choice of the appropriate measure depends on the situation at hand. The CV measure seems to be the more widely preferred, presumably because it addresses the more common situations and reflects a more widely held judgement concerning the just distribution of property rights.

4.3.2 Utility map and demand curve measures of consumer surplus

Conceptually useful as the notions of compensating and equivalent variation are for measuring welfare changes, they are not directly observable from market data. Unless we either make assumptions about the form of the utility function or employ questionnaire surveys, in practice it is necessary to measure consumer surplus in terms of the area under the demand curve and above the price line.[3] In this section we explore the relationships between the CV_1 and EV_1 measures of welfare change and the measure of consumer surplus beneath the demand curve. It is shown that the relevant measure of CV_1 and EV_1 following a price change is the appropriate area under the 'compensated' demand curve, and that if use is to be made of the ordinary Marshallian demand curve,

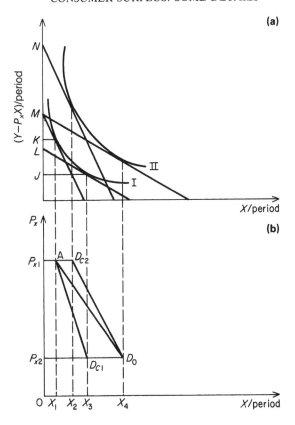

Figure 4.9 Consumer surplus and the demand curve: single consumer.

which is the one most readily identifiable in practice, it is necessary to
assume that the income effect of the price change is zero. In most
situations this is not thought greatly to distort reality.

Consider Figure 4.9 depicting a fall in the price of the commodity X, a
normal good. The ordinary Marshallian demand curve, holding money
income constant, is derived as AD_0 in Figure 4.9b. Now focus on
$CV_1 = LM$ and the derivation of the 'compensated' demand curve,
holding real income (or the level of welfare) constant (AD_{c1}). The total
payment at $P_{x1} = OP_{x1}AX_1$ (Fig. 4.9b) $= MK$ (Fig. 4.9a). The maximum additional amount the consumer would be willing to pay for the
price fall after removing the gain in real income created by the price
fall $= X_1AD_{c1}X_3$ (Fig. 4.9b) $= KJ$ (Fig. 4.9a). By addition, $OPX_1AD_{c1}X_3$
(Fig. 4.9b) $= MJ$ (Fig. 4.9a). Now, the total payment at P_{x2} given removal
of the consumer's additional real income created by the price fall $=$
$OP_{x2}D_{c1}X_3$ (Fig. 4.9b) $= LJ$ (Fig. 4.9a). So, by subtraction, $P_{x2}P_{x1}AD_{c1}$
(Fig. 4.9b) $= LM$ (Fig. 4.9a) $= CV_1$. Thus CV_1 is measured as the
additional area above the price line under the 'compensated' demand

curve (AD_{c1}). In similar fashion, EV_1 may be shown to be measured as the appropriate area under the other 'compensated' demand curve $(D_{c2}D_0)$. Only if the income effect is zero do all three demand curves coincide.

Thus precise measurement of the change in welfare requires evaluation of the area under a 'compensated' demand curve, whichever surplus measure is chosen as the more appropriate in the circumstances; and use of the Marshallian curve requires the assumption that the income effect is zero. Otherwise, the Marshallian measure of consumer surplus either under- or overstates the true change in welfare. To the extent that the proposal is relatively small scale, the assumption of a zero income effect may not be unreasonable. Indeed, Willig (1976) argues that in most situations the income effect is unlikely to be of practical significance. However, the analyst should be on guard against the possible bias involved in using readily observable demand functions.[4]

4.4 PRODUCER SURPLUS: SOME DETAILS

The same considerations as apply to the measurement of consumer surplus also apply to the measurement of producer surplus or economic rent. That is to say, changes in welfare should be measured as the area above the 'compensated' supply curve below the price line, or if the ordinary supply curve is used, it is necessary to assume that the income (or wealth) effect is zero. This is illustrated in Figure 4.10 for the case of an individual unit of a factor of production, in particular, labour. Results may be aggregated across individual units as well as factors to measure aggregate surplus.

In Figure 4.10a money income per period is plotted on the vertical axis; hours of work per period on the abscissa. In Figure 4.10b the hourly wage rate is plotted on the vertical axis. The slopes of rays from the origin in Figure 4.10a represent the hourly wage rate with $W_2 > W_1$. The indifference curves in Figure 4.10a are upward-sloping for the reason that work and money income are complements, their slopes increasing because greater and greater increments of income are required to induce the worker to give up successive amounts of leisure.

The ordinary supply curve is derived in Figure 4.10b as AS_0 and the 'compensated' supply curve based on the compensating variation as AS_c. So long as leisure is a normal good, AS_c lies below AS_0 giving a higher value of economic rent ($W_1AS_cW_2$ as compared with $W_1AS_0W_2$). $A'S_0$ is derived as the 'compensated' curve based on the equivalent variation and lies above AS_0 giving a lower value of economic rent ($W_1A'S_0W_2$). Only in the event that the income or wealth effect is zero do all three curves coincide. Use of the ordinary supply curve, therefore, as the basis for measuring economic rent requires the assumption of a zero wealth effect.

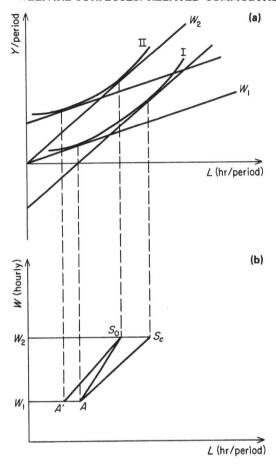

Figure 4.10   Economic rent and the factor supply curve.

4.5   WELFARE SURPLUSES: RELATED COMMODITIES

Having outlined the main principles of surplus measurement, it may be useful to make reference to the treatment of welfare surplus on goods or services other than the ones directly under analysis. As Mishan (1982a, Ch. 7) makes plain, there is no cause to include as benefit any change in consumer surplus in respect of a related good or service the price of which is not altered by the introduction or expansion in the level of output (and hence fall in the price) of the good or service in question. If, for example, a fall in the price of a good under analysis shifts the demand curve for a substitute good to the left, and thereby reduces consumer surplus attaching to the substitute good, this reduction is not to be counted. Provided that the supply price of the substitute remains constant, the reduction in consumer surplus is simply the consequence of consumers improving their welfare levels by switching from the substitute to the now relatively cheaper first good; and this welfare gain is measured in terms of the increase in consumer surplus on the first good. The possibility of a change in economic rent in connection with the second good does not arise.

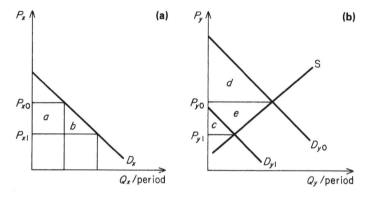

Figure 4.11   Welfare surplus changes: related commodities.

If, on the other hand, the prices of related goods or services are altered by the introduction or change in the level of provision (i.e. of price) of the first good, then it is necessary in principle to measure the change in welfare surpluses on related goods. Suppose the price of one good (X) falls from P_{x0} to P_{x1} as indicated in Figure 4.11a. The gain in consumer surplus is given by the area ($a + b$). The fall in the price of X in turn causes the demand curve for a substitute good (Y) to shift from D_{y0} to D_{y1} as shown in Figure 4.11b, giving a gain in consumer surplus of (c). The area (d), which may appear to be a loss of consumer surplus on Y, is not to be reckoned as a loss on the argument already outlined that it merely reflects the betterment of consumers as they switch to good X. As far as producer surplus is concerned, there is a loss on good Y of ($c + e$). The net welfare loss on good Y is, therefore, (e).[5] Thus the value of the fall in the price of X is measured as ($a + b - e$). The logic is readily extended to the cases of complementary goods. It is to be noted, however, that in many circumstances effects in respect of related goods may not be considered sufficiently significant to worry about. At the same time, they receive frequent attention in connection with transportation projects, one of the applications examined in Part II of the book (Ch. 10). Development of one transportation facility or mode can cause a switch of travellers from substitute, and to complementary, facilities or modes.

The foregoing discussion does not address an additional complication regarding the aggregation of surpluses on related goods when good X has an upward-sloping supply curve. This is the so-called 'path dependency' problem. When the price of good X is altered (reduced, in our example) by the project in question, the shift in the demand curve for Y and the associated fall in P_y in turn causes a shift in the demand curve for X, leading to a series of demand curve shifts for both goods. By way of example, X may be journeys between two places made by public transport; Y may be journeys by private car.

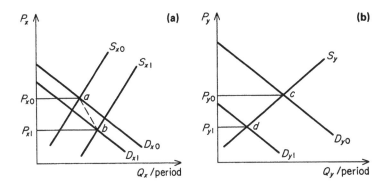

Figure 4.12   Path dependency: welfare changes.

Suppose final equilibrium obtains at P_{x1} and P_{y1} in Figure 4.12. How do we measure the aggregate welfare effect? On the one hand, addition of areas between original and new prices and under original demand curves (D_{x0} and D_{y0}) would overstate the effect. On the other hand, addition of areas under the new demand curves (D_{x1} and D_{y1}) would understate the effect. As a compromise, we could add the area under the original curve for one of the commodities to the area under the new curve for the other. This assumes that the path of adjustment involves first a fall in the price of one of the commodities (in this example commodity X) to its new equilibrium level without a shift in the demand curve, followed by a shift in the demand curve of the other commodity (Y) and a subsequent fall in its price. Alternatively, we could use intermediate areas, say $P_{x0}abP_{x1}$ and $P_{y0}cdP_{y1}$, assuming that the prices adjust at the same proportionate rate.

It can be shown (Pearce & Nash 1981, pp. 103–4) that these different measures of the aggregate welfare effect are equal to one another if cross-price derivates between all pairs of commodities are equal, a condition which holds in respect of 'compensated' demand curves or, otherwise, only in the case of homothetic utility functions in which the income elasticity of demand is unitary (Silberberg 1978, Ch. 11). Choice of any single measure of aggregate surpluses, therefore, involves the assumptions that zero income effects exist or else income elasticities are unitary,

assumptions which may not be greatly at odds with reality in many
circumstances.

4.6 SUMMARY

In circumstances where prices change in response to an initiative under
analysis, benefits are measured in terms of the sum of consumer and
producer surpluses. These are then compared with the capital and
continuing opportunity costs of the initiative. 'Compensated' demand and
factor supply curves should be used as the bases of surplus measurement,
or else the assumption has to be made that income (or wealth) effects are
negligible. Welfare surpluses in turn can be estimated in terms of either
compensating or equivalent variations (or surpluses), depending on which
measure is deemed the more appropriate, or tractable, in practice.
Welfare surpluses on related goods and services are to be included in the
net benefits only to the extent that the prices of related commodities are
affected by the initiative under analysis. The path dependency problem
may then be encountered. In practice, however, indirect effects may not
be quantitatively significant, or assumptions regarding the necessary
conditions of path independence may not be unreasonable.

NOTES

1 Strictly speaking, so-called 'compensated' demand and supply curves should be
   used. We leave this matter to Sections 4.3 and 4.4.
2 We address the issue of measuring the value of 'intangibles' such as 'peace and
   quiet' in Chapter 5.
3 For illustrations of the use of assumed utility functions and questionnaire
   surveys for estimating utility map measures of consumer surplus in the context
   of urban renewal projects, see Chapter 9.
4 Given a clear choice between CV and EV, of course, and given knowledge
   about whether or not the commodity is normal, the direction – if not the
   extent – of the bias can be known. This may be sufficient information on
   which to base a sound decision. On the other hand, procedures are being
   developed for deriving exact measures of consumer surplus from observable
   demand functions (e.g. Hausman 1981, Johnson 1985).
5 This demonstration is consistent with the material in Porter 1979 and Varian
   1979. In so far as it incorporates producer surplus into the measure of welfare
   change, it differs from Mishan 1982a, Ch. 8.

5

Measurement of benefits
and costs:
efficiency pricing

In Chapter 4 we discussed the general principles concerning measurement of economic and social benefits and costs in situations in which projects alter market prices. We now address issues of measurement in situations in which prices, whether they are altered or not, are deemed not to reflect true economic or social value and in which observed prices do not even exist. Market prices may fail to reflect true value in the presence of market distortions, disequilibrium and the relevance of non-efficiency considerations. Examples of the absence of market prices include services supplied publicly at zero price, and externalities such as noise or other pollution which are not exchanged in the market. In these situations we introduce accounting, or shadow, prices as the bases on which to form valuations of benefits or costs.

The concept of a shadow price derives from mathematical programming as the marginal value imputed to an input or output at the optimum. More generally interpreted in CBA, it refers to the value associated with a change in social welfare following use or loss of the marginal unit of an input or output. Shadow prices thus reflect true economic or social value at the margin. In this chapter we outline the need for shadow price adjustment when observed prices fall short of this standard, as a result of market distortions or the existence of disequilibrium, and when market prices are not available at all. Shadow pricing on these grounds is termed efficiency pricing (see Ch. 1). Adjustment of prices in connection with the relevance of non-efficiency considerations (social pricing) is deferred to Chapter 6.

Two caveats are entered at the start of the discussion. First, it is, strictly speaking, unsound to adjust the prices of market items on a piecemeal basis, for this involves inconsistencies unless the full range of non-optimal prices is also adjusted. For example, the shadow pricing of labour input to a project may also require the shadow pricing of other inputs, say steel and electricity, to the extent that labour is involved in their production. The difficulty is that information concerning all relevant input interrelationships is unlikely to be available. The analysis proceeds, therefore, on the assumption that distortions arising from omitted adjustments are not serious. Secondly, the task of measuring any shadow price may be so hazardous, as we shall see, that the very wisdom of the exercise might be questioned. It has been pointed out by McKean (1968) that the results of the effort to shadow price may be no more reliable as indications of true economic value than are observed market prices. Hence the analyst should give serious consideration to the balance of expected benefits and costs before engaging in shadow pricing. With these caveats in mind, we examine the question of efficiency pricing for market and non-market items in turn.

5.1   EFFICIENCY PRICING: MARKET ITEMS

In this section we examine the circumstances in which it is necessary to shadow price benefits or costs where imperfect markets prevail, indirect taxes or subsidies apply, commodities are traded internationally and unemployment of resources obtains. Underpinning the discussion is the general principle that the need to adjust market prices depends on the anticipated impact on, or adjustments in, the rest of the economy when the project is undertaken (Layard 1972, p. 18). In particular, it depends on whether or not inputs come from alternative uses and whether or not outputs displace other outputs. Also underpinning the discussion is the assumption that projects cause merely small changes in output so that we can, for simplicity, abstract from changes in surpluses, and value effects in terms of prices, suitably adjusted.

5.1.1   Monopoly imperfections

In the case of a project input (e.g. steel) being purchased in a less than perfectly competitive market such that $P > MC$, it is necessary to determine whether to value the input at its observed price as paid in the market or at marginal cost, that is, the cost at the margin of dedicating resources to its production. The case is illustrated in Figure 5.1 where the profit-maximizing monopolist restricts output to Q_0, forcing price up to P_0. In a perfectly competitive market Q_1 and P_1 would obtain.

Appropriate valuation of input cost depends on the principle enunciated earlier. If it is expected that production of steel will not rise to meet the additional demand created by the project, then purchase of the steel for the project displaces an alternative use for the steel and it should be valued at market price (P_0), the price which the alternative user would be willing to pay to acquire it. This represents the worth of the steel in its next best use and no shadow price adjustment is required.

If, on the other hand, it is expected that steel production will increase

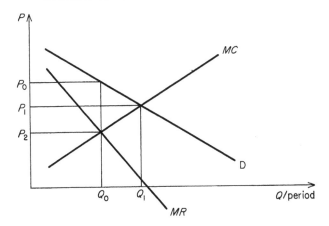

Figure 5.1 Monopoly market.

to meet the full requirements of the project (that steel will come from new supplies), then no alternative use is pre-empted and the steel is to be valued at marginal cost (P_2), the cost to society of allocating resources to the production of increased supplies. Thus a shadow price adjustment is used. If it is expected that a proportion of the steel requirement will pre-empt alternative use and another proportion be met from increased supplies, then a weighted average of price and marginal cost represents the appropriate shadow price.[1]

In the case of output from a project being sold in an imperfect market, analogous considerations apply. If the output displaces alternative output its value is measured in terms of the value of resources released from the alternative use, their marginal cost. If, on the other hand, it adds to market supply without displacing alternative output, it is valued in terms of its market price, the amount which consumers are willing to pay per unit (McKean 1968). In practice, of course, information requirements for shadow pricing in imperfect markets are so heavy that market prices are typically used.

5.1.2 Indirect taxes and subsidies

It is not necessary to discuss both of these cases in detail since subsidies may be viewed as negative taxes with exactly opposite effects. The problem for CBA is that indirect taxes and subsidies drive a wedge between resource cost and market price.

Looking at indirect taxes as they are attached to inputs, inputs are valued at market price if it is anticipated that additional supplies will not be made available once the project absorbs its requirements. If, on the other hand, additional supplies equal to the full amount of inputs used by the project can be anticipated, inputs are valued at marginal resource cost. A weighted average of price and marginal cost applies for the expectation of partial replacement of supplies, the weights being proportions of supplies replaced and not replaced.

In the case of project outputs subject to tax, the principle corresponds to that outlined for market imperfections. If the output displaces alternative output, its value is measured in terms of the marginal cost of resources released from alternative production. If the output adds to market supply without displacing alternative output it is valued in terms of market price.

As a final note, it is stressed that the case for shadow pricing in the presence of an indirect tax assumes that the sole purpose of the tax is to raise revenue. If the purpose is to correct externalities, as it may also be with a subsidy, there may be no need for adjustment since the tax or subsidy serves to align private and social cost.

5.1.3 Trade restrictions

In the case of inputs purchased abroad and outputs sold abroad, it is important to be clear about efficiency price. In particular, the question arises whether observed domestic price or world (border) price should be used when trade restrictions (tariffs, quotas or exchange controls) drive a wedge between the two prices.

In the absence of trade restrictions, and abstracting from indirect taxes, imported commodities sell on the domestic market at a price equal to border price converted at the equilibrium exchange rate. The unit domestic value of exports is also defined in terms of border price at the equilibrium exchange rate. But if trade restrictions maintain the domestic currency at an overvalued level (or foreign exchange at an undervalued level) so that the official exchange rate (OER) is below equilibrium (as in Fig. 5.2), the domestic value of traded commodities is higher than border price converted at the OER. As Figure 5.2 shows, willingness-to-pay for foreign exchange at the margin is SER (the shadow price of foreign exchange). Since SER is the price that people are willing to pay for the foreign exchange needed to acquire imports or made available as a result of exports, domestic value is equal to border price converted at the SER.[2] Thus, instead of valuing traded commodities at border prices converted at the OER, efficiency pricing requires that border prices converted at the SER be used (UNIDO 1972). This is tantamount to valuing these commodities at domestic market price.

As an alternative to the above procedure, it is possible to value traded commodities in terms of border prices at the OER and to shadow price non-traded commodities (adjusting them downwards by the ratio OER/SER) in order to maintain correct relative prices between traded and non-traded commodities (Little & Mirlees 1974). In Chapter 16 we demonstrate the equivalence of the two procedures.

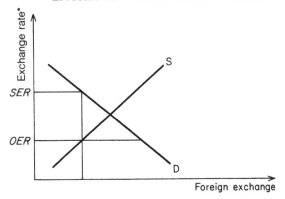

* Exchange rate = foreign currency in terms of domestic currency.

Figure 5.2   Foreign exchange market.

5.1.4   Unemployed resources

Where resources which would have been otherwise unemployed are employed on a project, there is a further example of the need for shadow pricing because alternative use is not precluded. We focus on the case of labour, although the principles involved can be generalized to other factors. At a basic level, the economic cost of labour, if it would have otherwise remained unemployed, is no more than labour's valuation of its leisure time now forgone. In Figure 5.3 the supply curve of labour can be interpreted as a curve showing the marginal value of leisure. Thus, if the prevailing wage rate is W_3, giving involuntary unemployment of L_1-L_3, the cost of taking on labour from the idle pool is only W_1, assuming that jobs are acquired by those workers having the lowest valuation of leisure. In practice, however, jobs are not necessarily allocated in this fashion so that the shadow price of labour from the idle pool is represented by some average valuation of leisure for this labour.

To the extent that labour is drawn from employment elsewhere, the situation involves some complication. If the labour in question is replaced elsewhere by other labour from the idle pool, its economic cost is again the average valuation of leisure for unemployed labour. If, on the other hand, labour used on the project displaces labour in other uses which cannot be replaced, then its social cost is based on the market wage. It should be clear that this represents the underlying principle of shadow pricing outlined earlier.[3]

Abstracting from complications, the shadow price (SP_i) of a unit of labour of type i employed on a project may, therefore, be written in the following form:

$$SP_i = (1 - r_i)W_{mi} + r_iW_{li} \qquad (5.1)$$

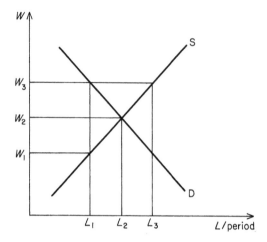

Figure 5.3 Labour market.

where W_{mi} = market wage of labour of type i, W_{li} = average valuation of leisure by labour of type i, and r_i = probability of drawing labour of type i from the idle pool.

Use of the market wage assumes that labour is paid the value of its marginal product. Estimation of the other terms in Equation 5.1 poses certain difficulties. One approach to the valuation of leisure is to observe actual behaviour in which people expend money to save leisure time; another is to ask people to express a value (see Section 5.2 below). In practice, leisure is often valued in terms of unemployment support payments, which when netted out as transfers (see below) leave $W_{li} = 0$. So far as measurement of the probability (r_i) is concerned, a direct survey is ideally required, although this may not reveal what alternative state labour would have been in had it not been employed on the project. Examples of surveys used to estimate probabilities (Evans 1973, Hodge 1982) identify previous rather than counterfactual labour states and are, therefore, less than fully satisfactory.

As an alternative procedure for estimating the probability that labour employed on the project under analysis would have remained otherwise unemployed, the use of 'synthetic response functions' has been suggested (Haveman & Krutilla 1968). If we plot the relationship between unemployment for labour of type i (ideally defined to include discouraged workers as well as the registered unemployed) and the probability of drawing the factor from the idle pool, the curve will rise from left to right, as in Figure 5.4. Given information on the unemployment rate, this curve may be used to estimate the probability (r_i) that labour is drawn from the idle pool. The problem is to define the precise shape of the curve. Haveman and Krutilla suggest that sine functions may provide an appropriate depiction of response curves; they further suggest that a range of values for r_i be established relative to each rate of unemployment. If r_i is likely to approach unity at, say, 20–25% unemployment, and full employment is defined as any rate up to, say, 6% unemployment, a range of likely values for r_i is established at each unemployment rate. At 10% unemployment, for example, the probability $(1 - r_i)$ of drawing labour from alternative employment would be between 0.50 and 0.85. Both figures would be used to establish a range of values for labour's shadow price. Approximation through sensitivity analysis in this way, it is suggested, is about as much as we can really hope to achieve in the circumstances. Haveman and Krutilla provide extensive illustrations of the application of this approach in connection with capital as well as labour inputs.

Refinements to the basic model of shadow pricing in the presence of unemployment may be introduced as appropriate. Thus account may be taken of various costs that require to be added to or subtracted from forgone leisure costs in identifying the full shadow price of labour employed on a project. Examples include labour migration costs either incurred or averted as a result of job creation on the project (Little & Mirlees 1974, Jenkins & Kuo 1978, Boadway & Flatters 1981). Incremental public infrastructure and public service operating costs associated with induced or averted labour migration (Little & Mirlees 1974) may also be recognized, as may labour turnover adjustment costs incurred by employers as labour moves out of other employment into jobs created by the project (Hodge 1982). It is to be noted as well that unemployment support payments, lost by newly employed workers but saved by the rest of society, are to be netted out as transfers in so far as they are included in the value of forgone leisure (see Mishan 1982a, Ch. 11).[4]

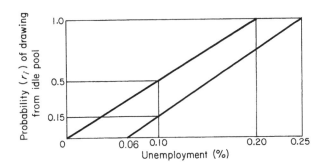

Figure 5.4 Synthetic response functions.

5.2   EFFICIENCY PRICING: NON-MARKET ITEMS (INTANGIBLES)

In addition to situations in which market prices are inappropriate as measures of true economic cost or benefit, it is necessary to generate shadow efficiency prices for items of benefit and cost which lack market values altogether. Such items include government-supplied commodities which usually have characteristics of publicness about them (non-exclusion and non-rivalry), externalities (e.g. air, water and noise pollution) and such other commodities as time and life which are not exchanged explicitly in the market. Transport improvement represents a good example of a type of project which may generate benefits in terms of both time and life savings. In what follows we outline alternative approaches to the measurement of these intangibles.

5.2.1   Contingent valuation

It is possible to establish the value of some beneficial contingency (say, pollution abatement) by asking people directly what they are willing to pay at most for it or how much they require at minimum by way of compensation if they are to be denied it.[5] Similarly, maximum willingness-to-pay to avoid a cost contingency, or minimum compensation required to suffer it, may also be established. The procedure may involve simply asking respondents to state a single value. But a more sophisticated version of the method engages respondents in a bidding game in which they express willingness-to-pay at successively higher prices or need for compensation at successively lower prices, the process being continued until final bids are established reflecting maximum willingness-to-pay or minimum compensation required.

A problem with the contingent valuation method is that various sources of bias discussed below may influence responses and weaken the approach if enquiries are not carefully designed. However, recent research (e.g. Brookshire & Crocker 1981, Schulze et al. 1981, Brookshire et al. 1982) suggests that concern over these biases may have been exaggerated in the past, and there is developing a growing acceptance of the method, particularly for estimating the value of environmental amenities. Examples of its application in the field of urban and regional planning are discussed in Part II of the book.

So-called strategic bias is a leading cause of concern, being present to the extent that responses are conditioned by expectations regarding charging arrangements. If the respondent anticipates that his/her response will be used as the basis for financing the provision of the scheme under investigation, there may be an incentive to understate the willingness-to-pay or necessary compensation. If he/she expects that the good or service will be provided free, the incentive may be to overstate true value. As a way around this difficulty, it has been proposed that no information regarding charging be given so that either true preferences will be revealed or individual biases will tend to cancel out (Bohm 1971). Alternatively, respondents may be divided into two separate groups and provided with contrasting information regarding charging plans. It may be that upper- and lower-bound values will then be generated which can be used as the basis for sensitivity analysis.

Other sources of concern include hypothetical, information and instrument biases. Hypothetical bias arises out of the hypothetical situation in which the respondent is placed. Thus he/she may not be able to imagine the environment postulated in the questionnaire or bidding game and accordingly may not be able to provide a meaningful response. Additionally, the absence of budget and time constraints in the hypothetical situation may bias responses. Information bias involves the possibility of variation in responses, depending upon information presented. In so far as charging arrangements are concerned, this bias overlaps with strategic bias. But it also encompasses other situations where lack of, or type of, information conveyed to respondents may influence expressed valuations. Willingness-to-pay for environmental clean-up, for example, may depend on the period of time the clean-up operation will take, on average expenditures already being incurred on clean-up and on the payment vehicle to be used. Thus it is important not to omit presentation of relevant information on these matters. Moreover, the process of questioning may give rise to instrument bias. In the context of a bidding game, for example, valuations may depend on the price at which respondents are asked to start bidding. It is, therefore, advisable to choose starting bids at random. Finally, sampling, interviewer and non-respondent biases may also infiltrate the survey approach.

Clearly, care is required in using the contingent market method. At the same time its advantages are to be recognized (Brookshire & Crocker 1981). First, it allows the analyst to take account of phenomena not in the range of historical experience. Thus, for example, valuation of the threat of air pollution from a proposed industrial development may be attempted. Secondly, the method is useful for separating effects which other methods discussed below may not be capable of separating. For example, the aesthetic and health impacts of a pollution abatement scheme may be distinguished, a distinction not possible with the hedonic price method discussed below. Thirdly, estimation of demand or supply functions by use of the contingent valuation method permits evaluation of surpluses which are not always identifiable using procedures discussed below.

5.2.2 Behaviour observation

Rather than rely on what people say, economists often prefer to observe actual behaviour and from it to infer implicit valuations for items to be measured. Leading examples of this approach include transportation mode, route and speed choices as a basis for valuing non-working time; the method developed by Clawson (1959) and Clawson and Knetsch (1966) for estimating the demand curve for, and hence value of a recreational facility; the hedonic price method for valuing both pollution (or pollution abatement) and the statistical value of life; and the use of so-called 'defensive' expenditures as implicit indicators of the value of reducing the effect of such 'bads' as pollution and risk. These methods are covered more fully in the context of their use in urban and regional planning exercises in Part II of the book. Here we provide a brief outline of each.

Time valuation on the basis of transportation choices involves observing how much extra money people are willing to pay to travel by a faster mode (e.g. Concorde over regular flights), a faster route (e.g. motorways over regular roads) or at a faster speed on a given mode or route (e.g. by car on the motorway). For good summaries of variations on this theme, see Harrison and Quarmby (1969) and Watson (1974). It is stressed that this approach to determination of willingness-to-pay tends to understate true value, in so far as consumers' surplus is not captured in observed actual payments.

The Clawson–Knetsch approach to the valuation of a recreational facility (wilderness or urban park, swimming pool, theatre, etc.) involves estimation of the demand curve for the facility from travel cost data. Since integration over the relevant area of the demand curve is made possible, this method does capture consumers' surplus. Briefly, the method is based on the relationship between visits per period and distance travelled. From this negative relationship can be constructed a downward-sloping cost/visit curve and from that a demand relation based on hypothetical entrance charges. For clear and simple examples of the application of the method, see Smith (1971) and Mansfield (1971). Refinements to the method continue (see Ch. 11).

The hedonic price approach comprises econometric estimation of the relationship between the value of a thing and its various attributes. In general form, the function to be estimated is therefore:

$$V = V(A_i \ldots A_n) \tag{5.2}$$

where V = value, A = vector of attributes, and the marginal valuation, or hedonic price, of the ith attribute is given as $\partial V/\partial A_i$. Most widespread use of this method is for valuing pollution effects using house prices as the dependent variable and structural and locational features of homes, including environmental quality, as independent variables (see, for example, Ridker & Henning 1967, Anderson & Crocker 1971). The estimated coefficient(s) on the environmental quality variable(s) provide(s) the implicit price(s) of pollution per house affected. The method has also been used to place a value on life, using (say, weekly) wage rates (ω) as the dependent variable with characteristics of jobs, including risk of death, as independent variables (see, for example, Thaler & Rosen 1976). Thus the value of a life (or death avoided) is given as:

$$\sum_{i} WTP_i/n \; \mathrm{d}P = \bar{W}\bar{T}\bar{P} \; \mathrm{d}P \tag{5.3}$$

where WTP_i = willingness-to-pay by the ith individual for a specified change in the probability of death ($\mathrm{d}P$), $n \; \mathrm{d}P$ = expected lives saved, and $\bar{W}\bar{T}\bar{P}$ = mean willingness-to-pay for $\mathrm{d}P$. $\bar{W}\bar{T}\bar{P}$ is estimated from the coefficient ($\partial\omega/\partial P$) on the risk variable which provides a measure of the reward that workers are willing to give up per week (their willingness-to-pay) for decreased exposure to risk at work.

The hedonic price approach assumes full information regarding the attributes of different locations and jobs. It also assumes freedom to move between different locations and jobs. Moreover, it encounters the standard econometric problems of model specification, data reliability and data availability.[6]

A final method involving behaviour observation is to impute a value for certain 'bads' from expenditures designed to protect society from these misfortunes. This may be called the 'defensive expenditures' method. An example places a value on noise nuisance by observing expenditures on sound-proofing of homes (Starkie & Johnson 1975). Other possibilities might be to value the cost of death and injury on the basis of observed outlays on such items as motorway crash barriers, search and rescue operations and medical research. A problem with this method is that expenditure decisions are made in light of considerations other than merely defence against the 'bad' in question (e.g. double-glazing of house windows conserves heat and protects against noise), so that a considerable amount of statistical 'noise' tends to infiltrate estimates.

5.2.3 Other market proxies

Another approach to the estimation of intangible benefits and costs is to use dollar values generated in the market in connection with activities other than the defensive purposes referred to in the last section. Thus the value of a service provided free of charge through the public sector (e.g. the ambulance service) may be approximated by reference to the price charged for the same (or similar) service, if available, in the private sector. Such a procedure obviously understates full value to the extent that consumers' surplus is not captured; it also involves problems of comparability so far as commodities are concerned.

Cost savings may also be used as a measure of project benefits. For example, the value of domiciliary health care may be gauged in terms of the cost of alternative residential care avoided, although this approach abstracts from the likelihood that the two alternatives yield differential psychic benefits to patients and/or friends and relatives (see Ch. 13). Part of the benefit of a preventive health programme may be gauged in terms of treatment costs avoided; part of the benefits of pollution abatement come in the ` form of avoided damage costs to buildings and other structures; part of the benefits of a road improvement scheme come in terms of travel cost savings to people who would have travelled despite the improvement scheme.

Finally, the value of working time saved (as a result of an improved transportation link) may be approximated by use of average wage rates (see Watson 1974); and the value of life or of human capital programmes in the fields of health, migration, education and training by use of lifetime earnings. In the case of human capital programmes, benefits are expressed in terms of incremental earnings attributable to programmes (e.g. Doessel 1978, Wilson 1980). Obvious difficulties with the earnings approach are that it abstracts from external impacts on other parties (except as they benefit through tax revenues levied on incremental earnings) and implies that programmes yield no benefits to non-earners. The latter implication is that life carries no value for its own sake. In Part II of the book we review the use of market proxies along with contingent valuation and behaviour observation methods as they have been used in analyses of programmes in the field of urban and regional planning.

5.2.4 Additional approaches for dealing with intangibles

Measurement of intangibles based on contingent valuation, behaviour observation or the market proxy method is often, of course, open to dispute. It may be considered more realistic in many circumstances to adopt a less ambitious procedure which does not involve attempts to put a monetary value on all effects identified as relevant. Three such procedures are outlined here: the cost-effectiveness, the decision matrix and the qualitative assessment approaches.[7]

Cost effectiveness may be used in circumstances where monetary values for cost are available while benefits are not considered to be capable of reliable measurement in monetary terms. However, physical measures of benefit are available. In these circumstances, it is possible to analyse proposals from the point of view of cost minimization, given a specified objective, or of benefit maximization, given a specified budget. For example, analysis of crime prevention programmes may identify the least cost way of reducing the incidence of a certain type of crime by a given number of crimes per annum, or the best way of spending a given budget on crime prevention, measuring benefits in terms of reductions in the crime rate.

Decision matrices provide information which helps decision makers to focus on critical values for intangible items without providing direct measurement of their monetary value. It can be shown, for example, what will be sacrificed in terms of forgone income from forest or mineral production if scenic amenities are preserved. Decision makers are then left to decide whether the implicit trade-off between scenic amenity and income is worthwhile. This approach amounts to presenting a 'switching' value for intangibles which will tip a measured net benefit in the other direction. For comparison of alternative projects a decision matrix may be presented showing the trade-off between impacts that are measurable in monetary terms and impacts that are measured in terms of physical units. Thus, if the matrix appears as in Table 5.1, decision makers must decide whether the extra $10 of monetary net benefit associated with project A is worth the sacrifice of four units of non-monetary benefit.[8] Although the analysis does not provide a cut-and-dried solution, it narrows the 'zone of ignorance' regarding proposals and focuses attention on the implicit valuations which are embedded in decisions. In this way, it provides a sounder basis for decision making than would exist in its absence.

Table 5.1 Decision matrix

Projects	Net benefits	
	Monetary	Non-monetary
A	$100	11 units
B	$ 90	15 units

Finally, where it is possible to generate neither physical nor monetary measures of benefit or cost, it is usually feasible to list the important intangible effects and perhaps to provide some indication as to their relative significance. Even the provision of such minimum qualitative information is useful for decision making.

5.3 SUMMARY

Shadow efficiency prices are designed to reflect true economic value. Market items may require to be shadow priced in circumstances involving imperfect markets, indirect taxes and subsidies, internationally traded commodities, and factor unemployment. The guiding principle concerning the need for efficiency prices is that adjustment of market prices is required to the extent that inputs are not displaced from alternative use and outputs do displace alternative output.

Efficiency pricing is further required when items of benefit or cost lack market values altogether. Alternative approaches to the measurement of such intangibles include the use of questionnaire surveys or bidding games (the contingent valuation approach), direct observation of human behaviour for the purpose of eliciting implicit valuations, market proxies and a selection of other approaches which avoid the complete monetariz-ation of all benefits and costs. The latter include the cost-effectiveness method whereby dollar values are employed on the cost side and physical unit measures on the benefit side of the account, and decision matrices in which critical trade-offs between monetary and non-monetary effects are highlighted. The mere provision of purely qualitative information regarding benefits and costs may also be useful.

NOTES

1 The same logic applies in the allied case of the factor of production, labour, when its price is maintained at an artificially high level due to forceful collective bargaining or the existence of a minimum wage above equilibrium.
2 Note that this analysis assumes that foreign exchange is allocated to those making the highest bids. If this is not the case, as is likely, then an average value above *OER* represents the economic value of a unit of foreign exchange.
3 An exception to the need to shadow price even when factors would have remained otherwise unemployed occurs when a target unemployment policy is being pursued such that additional employment through the project leads to reduced employment elsewhere in the economy and the cost of factors employed on the project is the forgone output elsewhere, as measured by observed factor price. Similar considerations apply in respect of the hypothesis of the 'natural' rate of unemployment. If the economy is at its 'natural' rate, any job creation on a project will be offset eventually by job losses elsewhere.
4 If an unemployed worker in receipt of $200 per week in unemployment benefit would be willing to take a job for at least $200 per week (the value of his forgone leisure) and he is paid that amount, the economic cost of employing him is zero. The reason is that he is no worse off than previously, nor is the rest of society. Similarly, if the value of his leisure time is $250 per week so that he would not accept a job for less than that amount, his shadow price is $50 per week.
5 It is not to be expected that these two measures will give exactly the same answer, even if the income effect is zero. Empirical studies suggest maximum willingness-to-pay may be typically lower than minimum compensation (Knetsch & Sinden 1982). One explanation may be that the former is constrained by income and wealth, whereas the latter is not. Another explanation may be that marginal dollars may be more valuable when lost (paid out) than when gained (received as compensation).
6 For a considered discussion of the worth of the approach, see Freeman 1979.
7 A fourth procedure is referred to in Chapter 6. This is the differential

weighting of measured benefits and costs to reflect their relative significance in terms of intangible effects.

8 Extended versions of this approach are represented by multi-objective matrix display models outlined in Chapter 6.

6

Measurement of benefits and costs: distributional considerations

The focus on aggregate efficiency in economic CBA has been viewed as a major limitation of this traditional form of CBA (see, for example, Maass 1966). Distributional implications of projects, programmes or policies for, say, income classes, ethnic groups and regions are often of compelling interest to policy makers; and in lesser developed countries (see Ch. 16) the distribution of benefits between savings and consumption is also a pressing concern, given an extreme deficiency in savings as required to fund economic growth. For these reasons it is not surprising that there have been attempts to incorporate distributional considerations into what is now widely termed 'social' CBA.

In this chapter we address the issue of distribution between members of the present generation. The issue of distribution between generations is deferred until Chapter 8. Before proceeding, it should be emphasized that all manner of unmeasured intangible effects (e.g. environmental impacts, national security) may be dealt with along the lines discussed for distributional matters, even though such effects may in some cases be classified as efficiency effects.

6.1 GENERAL APPROACHES

There are, in principle, three general ways in which distributional (or other incommensurable) considerations may be integrated into decision making using CBA (Marglin 1962). First, decision makers may maximize aggregate efficiency subject to a constraint that minimum conditions be met regarding the distribution of income and wealth (or other incommensurable concerns). For example, in the regional context projects are deemed worthwhile in order of aggregate efficiency impact, providing that designated regions benefit to a specified extent, or are not harmed more than to a certain extent. It is clear that this approach reflects the Little criterion discussed in Chapter 2. The second approach reverses the first. Decision makers maximize a distributional goal (e.g. to favour certain regions), subject to an aggregate efficiency constraint (e.g. that the proposal at least breaks even in efficiency terms).

The third general approach to consideration of distributional matters is to maximize a multidimensional social welfare function (the Bergson criterion of Chapter 2). Decision makers maximize a weighted sum of net benefits accruing to different groups in society, the weights reflecting the relative social importance of creating a dollar of net benefit for different groups. Thus, in the regional context, a dollar of net benefit for a disadvantaged region would be weighted more heavily than one accruing to a more prosperous region. This procedure is known as social pricing.[1]

6.2 CONSTRAINT AND WEIGHT DETERMINATION

The problem, of course, in implementing the above general approaches lies in determining the value of constraints or weights. Several methods are available.

6.2.1 Suggested methods of constraint and weight determination

Since constraint or weight values reflect judgements regarding the relative importance of distributional and aggregate efficiency objectives, a first method is to leave the exercise of judgement to a chosen authority. Maass (1966) suggests that this responsibility should be entrusted to political institutions, a procedure which may not be feasible for every proposal considered, particularly when judgements will alter over time. But for relatively major undertakings it may work through, for example, the process of public hearings. Alternatives are to leave the responsibility to experts within the bureaucracy (subject to review by higher political authority; Freeman 1970) or to the analyst alone (Worswick 1972).

Other suggested methods relate specifically to the estimation of distributional weights rather than the establishment of constraint values. One is to base weights on some mechanical, if arbitrary, rule; for example, the inverse of the ratio of group mean income to population mean income (Foster 1966). Another method is to derive estimates of weights as implicit in social decisions already taken. Thus Eckstein (1961a) has suggested that, assuming that tax schedules reflect the principle of equimarginal sacrifice, the inverse of marginal income tax rates might form the basis of weight determination for different income classes. It might seem preferable to use the total tax system instead of income tax rates alone, and the assumption of equimarginal sacrifice might be questioned. Moreover, the judgements concerning equity as embedded in the income tax schedule may not correspond closely to the distributional purposes of expenditure policies. It is possible that a society may wish to encourage saving and work effort on the part of higher income groups through relatively low taxation rates while aggressively pursuing egalitarian designs through public expenditure initiatives. Examples of the application of this method, with particular reference to the regional distribution of project impacts, include an analysis of US Army Corps of Engineers programmes (Haveman 1965), and an analysis of alternative sites for a third London airport (Nwaneri 1970), both using tax-based regional weights.

An alternative to the use of tax schedules is to seek implicit weights in past public expenditure decisions (Weisbrod 1968). If analysis reveals that less efficient proposals were adopted ahead of more efficient ones because the former favoured certain groups, minimum estimates of the implicit distributional weights can be made, provided that there are as many income groups as proposals. Weisbrod illustrates the method with an example involving four past projects (1 to 4) yielding net benefits to four distinct groups of people (a to d). Project 1, the most efficient project, was not undertaken while the other projects were. Taking account of non-efficiency considerations, therefore, projects 2 to 4 must have been viewed by decision makers as at least just as desirable as project 1 in terms of the benefit–cost ratio.[2] Thus a system of benefit equations may be written as follows:

Project

$$1 \qquad \alpha_a B_{1a} + \alpha_b B_{1b} + \alpha_c B_{1c} + \alpha_d B_{1d} = B_1$$

$$2 \qquad \alpha_a B_{2a} + \alpha_b B_{2b} + \alpha_c B_{2c} + \alpha_d B_{2d} \geqslant B_1(C_2/C_1)$$

$$3 \qquad \alpha_a B_{3a} + \alpha_b B_{3b} + \alpha_c B_{3c} + \alpha_d B_{3d} \geqslant B_1(C_3/C_1)$$

$$4 \qquad \alpha_a B_{4a} + \alpha_b B_{4b} + \alpha_c B_{4c} + \alpha_d \beta_{4d} \geqslant B_1(C_4/C_1)$$

where the coefficients $\alpha_a \ldots \alpha_d$ are weights indicating the relative importance of a marginal dollar of benefit for each population group. Note that right-hand side benefits for projects 2 to 4 are adjusted for the difference in costs between these projects and project 1. If equality signs are now assumed in place of inequalities, simultaneous solution of the four equations gives minimum estimates of the implicit values of $\alpha_a \ldots \alpha_d$.

The obvious difficulties with Weisbrod's approach are that full information on the part of decision makers is assumed, that social preferences may change over time, that estimation errors may be large and that various political considerations aside from concern for distributional equity may have influenced the decision to adopt the less efficient proposals. Despite these limitations, the method can provide interesting information regarding the *de facto*, if not the intended, significance attached in past decisions to non-efficiency concerns. In this sense the approach may assist decision makers to establish weights for the current period. An example of its use for weight determination employs health care expenditures to derive income class weights (Neenan 1971).

As a fall-back, should the preceding approaches to weight estimation be considered somewhat cavalier, there is always sensitivity analysis for showing results according to a range of likely weights. Decision makers may then choose their preferred weight values. A good example of this procedure in the regional development context involves the evaluation of a highway development scheme using a range of weights greater than unity for benefits accruing to the depressed Appalachian region of the USA (McBride 1970).

6.2.2 Opposition to differential weights

With a myriad of difficulties clearly attending the development of distributional weights, it should scarcely be a surprise that there exists opposition to any attempt to introduce such weights. An influential body of opinion presents the following arguments against the use of differential weights. First, to ignore explicit consideration of distributional matters is simple, robust (it answers practical problems) and consistent with a long tradition in economics whereby the economist accepts that he is not professionally qualified to pronounce on non-economic matters (Harberger 1971). Moreover, to do otherwise leaves the door open to data manipulation in the sense of providing quantitative rationalizations for the pet projects of those currently in authority, removes the independence of the analyst *qua* economist, leads to inconsistencies over time in so far as weights change, and obscures the essential efficiency dimension of issues (Mishan 1974, 1982b). Differential weighting may also involve acceptance of larger losses of efficiency than people may realize, as well as carrying such surprising implications as higher project net benefits with higher factor payments and higher marginal income tax rates in lower income brackets (Harberger 1978). In addition, the efficient way to attend to distributional concerns, it is argued, is through tax/transfer policies, leaving public expenditure to be planned with a view to maximizing aggregate efficiency (Musgrave 1969). Notwithstanding the value of sensitivity analysis, the preference among this group of commentators is obviously to allow decision makers to make their own best judgements regarding the trade-off between efficiency and equity objectives in light of the results of economic CBA alone. Economic analysis will demonstrate the extent of efficiency to be sacrificed should a decision be made to pursue some distributional objective.

Another source of criticism of weight determination relates, not to the use of weights in principle, but to their *a priori* specification for purposes of analysis. This concern applies as well to the prior determination of constraint values. It has been argued that prior determination of precise point values that are given for the analysis may not make much sense if, as is often likely, there is no clear understanding on the part of decision makers regarding the structure of the social welfare function they are attempting to maximize (Cutt & Tydeman 1976). Without a clear idea of what objectives are really relevant, as well as their relative importance, it may be better to allow weights to emerge gradually through an iterative and interactive learning process, if they are to be used at all. Acting in the role of what has been termed the 'adaptive black box' (Cassidy & Kilminster 1975), decision makers would be asked to suggest at each iteration the direction of change in weights in order to improve the solution until an optimal set of weights is obtained. Although this method has been developed largely in connection with multi-objective programming models (e.g. Candler & Boehlje 1971), it has been illustrated in the context of CBA and the US Economic Development Administration programme of regional assistance (McGuire & Garn 1969). Its cumbersome nature, however, should not be overlooked from the practical point of view.

6.3   MULTI-OBJECTIVE MATRIX DISPLAY MODELS

An approach to the implementation of social CBA which has received relatively wide attention involves explicit disaggregation of the analysis according to different parties affected and according to different types of benefit and cost. This approach can deal with both monetary effects and incommensurable effects and can either include or exclude differential weights.

6.3.1   Examples of multi-objective accounts

It is now standard practice in the field of water resource analysis in the USA to use this method, studies of proposals being required to be undertaken from the separate points of view of different regions, different income classes and environmental impact as well as the aggregate point of view of impact on national income (US Water Resources Council 1973). It may be possible to consolidate results into a succinct summary matrix of the kind illustrated in Table 6.1. No attempt is made to measure every effect in monetary terms. In some cases of incommensurable effect there may be an alternative measure in physical units that can be used, while in other cases it may not be possible to do more than indicate the likely direction of impact. A single final decision criterion (e.g. net present value, benefit–cost ratio, internal rate of return) is not derived so that the analysis lacks the simple elegance of conventional CBA. On the other hand, the method is more compre-

Table 6.1 Multi-objective display matrix

		Benefits							Costs						
			National	Regional	Income class					National	Regional	Income class			
		Project effect	I+II+III (L+M+U)	I	II	III	L	M	U	Project effect	I+II+III (L+M+U)	I	II	III	L	M	U
Monetized effects	Real final goods and services																
	1	Sum real benefit								1	Sum real costs							
	2	Transfers								2	Transfers							
	3	Regional account								3	Regional account							
	4	Income distribution								4	Income distribution							
Non-monetized effects	Environmental	Aesthetics								Aesthetics							
		Important sites								Important sites							
		Water and air quality								Water and air quality							
		Irreversible consequences								Irreversible consequences							
		Other								Other							
	Non-environmental	Life								Life							
		Health								Health							
		Safety								Safety							
		Education								Education							
		Culture								Culture							
		Recreation								Recreation							
		Emergency preparation								Emergency preparation							
		Other								Other							

Note: line 3 = line 1 + line 2; line 4 = line 1 + line 2.
Source: Sassone and Schaffer, 1978, p. 172.

hensive, providing the decision maker with as much information as possible on which to base the judgements he then has to exercise concerning trade-offs between different objectives.

Prominent among matrix display models are two designs developed in the 1960s by planning experts: Lichfield's Planning Balance Sheet (PBS; e.g. Lichfield, 1964, 1966a,b, 1968, 1969, 1970) and Hill's Goals Achievement Matrix (GAM; Hill 1968). Variations on the theme have also been proposed (e.g. Hartle 1976, Poulton 1982). Here we merely outline the main framework of the PBS and GAM approaches. More complete details of matrix display models, along with case study examples, are presented in Chapter 12.

The PBS was developed for the purpose of evaluating urban and regional development plans having multisectoral impacts, although in principle the design can be applied to proposals for a single sector such as housing or transportation. The aim is to identify all relevant impacts (economic, social, environmental) on the urban or regional system, including the distribution of impacts between different actors in the system (classified as either producers/operators or consumers of goods and services). As Lichfield explains it:

> An initial step is to enumerate the sectors of the community which are affected by the alternative proposals, treating them on the one hand as producers and operators of the investment to be made in the system of new projects and on the other hand as consumers of the goods and services arising from those projects. Then for each sector the question is asked: 'What would be the difference in costs and benefits which would accrue under the respective schemes under examination?' The nature of the costs and benefits comprise all those which are of relevance to the planning decision. They thus include those which are direct as between the parties to the transaction and those which are indirect and come within the conventional definition of social costs and benefits; it includes those which relate to real resources and those which are transfers; it includes those which cannot be measured as well as those which can. Thus by means of certain specially designed tables, notation and rules it is possible to evolve and summarize a set of social accounts for each sector of the community, showing clearly the differences in costs and benefits which will accrue to them under the alternative plans. (Lichfield 1968, p. 16)

In its early formulations the method was criticized for not making explicit the objectives of the plan under scrutiny (Hill 1968). Costs and benefits, after all, cannot be defined except in terms of contributions to or detractions from programme objectives. The result is that there may have been a tendency in using the PBS to overlook the need to weigh the relative importance of different objectives in assessing results. In later formulations, objectives have been identified in PBS accounts (e.g. Lichfield & Chapman 1970).

Hill's GAM arrays cost and benefit items according to programme goals, and provides for explicit consideration of the importance of different objectives by attaching a relative weight to the effects listed under each goal. Weights are also attached to costs and benefits as they are estimated for different groups, these weights reflecting the community's desired distribution of effects. The approach was developed for use in connection with plans in the transportation sector, but there is no reason in principle why it should not apply to other types of proposal. It clearly runs counter to the case against prior specification of the objective function, although it can be used in conjunction with sensitivity analysis and/or interactive decision making. However, the practical problems of identifying all the weights precisely should not be minimized.

6.3.2 *Concerns regarding multi-account analysis*

Tested methods, then, can account simultaneously for multiple objectives. However, as a result of their multidimensional character, estimates are less simple to interpret than in the case of traditional economic CBA. In urban planning applications (e.g. the PBS and GAM), the approach also tends to obscure the efficiency dimension of issues, not simply because distribution effects are highlighted, but because applications tend not to measure impacts in terms of welfare surpluses, nor to involve efficiency pricing. Yet there is no reason in principle why models should not use conventional CBA concepts.

The foregoing points aside, multi-account analysis has run into opposition on other grounds. In the 1970s a group of water resource economists argued against incorporation of social, environmental and, most particularly, regional distribution effects into analyses of water resource projects, the area of application in which CBA had its main roots. We briefly outline and assess their arguments below.

Freeman & Haveman (1970) suggested that it was not important to recognize regional development as a concern of water resource planning since there was no evidence that plentiful water or the availability of inland waterway transport exerted a significant effect on regional growth. It was also argued that in order to determine whether benefits and costs attributed to any region were national efficiency effects or merely interregional transfers, it was necessary to measure impacts on all regions in the system, not merely a selected region, and this requirement presented enormous difficulties of measurement. On the other hand, however, projects in fields other than water resources may exert significant regional growth effects, and analysis of the impact of a project on selected regions may be sufficient for the purpose at hand. Moreover, the existence of data difficulties is no reason for not trying to complete a multidimensional analysis.

Another argument developed by Freeman & Haveman was that the 'only meaningful equity consideration is the distribution of income among individuals' (1970, p. 1537). The argument is that to focus on the distribution of income or welfare between regions may exacerbate inequity between individuals, since assistance to a disadvantaged region may substantially benefit the better off in both the region in question and, through interregional linkages, in other regions too. However, if it is widely considered to be important from the point of view of social justice to reduce interregional disparities in economic opportunity, then identification of regional distribution effects is justified.

It was also observed by Freeman & Haveman that the introduction of the regional development objective into water resource planning without introducing it into all policy choices would lead to invalid and costly sub-optimization in public expenditure planning. The danger was that potentially more effective means of stimulating regional development could be overlooked; for instance, manpower training, industrial subsidies or social overhead investment. Furthermore, recognition of regional development as an objective of water resource development could involve, they argued, surrender to the advocacy of special interest groups capable of making an equity case for all manner of inefficient proposals. However, while the dangers of sub-optimization and interest group pressure are to be guarded against, they are hardly unique to the consideration of regional distribution effects.

Finally, Freeman & Haveman argued that identification of regional benefits and costs is only relevant if well defined regional weights are developed and accepted, and it was their opinion that such weights were not capable of being generated. On the other hand, difficult as it may be to develop weights, we have seen that it may not be impossible, especially if sensitivity analysis is used; and, as we shall see in Chapter 16, it is standard practice in project appraisal in lesser developed countries to engage in social pricing based on weights generated in the bureaucracy. Furthermore, it does not follow from the difficulty of devising weights that distribution effects should not be identified at all. As discussed earlier, decision makers may be left to weigh the trade-offs between competing objectives from information on project impacts disaggregated according to groups.

Discussing the principles and standards of analysis as proposed by the US Water Resources Council (1970), Chicchetti et al. (1973) rehearse more or less the same points as Freeman & Haveman in coming to the conclusion that additional accounts to measure project impacts on regional development, environmental quality and social wellbeing are 'redundant and methodologically unsound' (p. 724). National economic development, these authors assert, should remain as the sole criterion of water resource development if deterioration in the quality of investment decisions is to be avoided.

Yet, for all the strength of conviction on the part of these commentators, work has been progressing steadily since the 1960s on the matter of integrating distribution effects into analyses. The reasons are obviously that regional equity and other distributional concerns remain persistent planning issues (not only in the water resources field), and that a number of the arguments advanced by the opponents of social CBA can be seen to be at least open to challenge as indicated above.

6.4 SUMMARY

It is clear that social as well as economic CBA is an important form of analysis, given that redistribution of income and wealth as well as such other non-economic goals as national security enter the social welfare function along with aggregate efficiency. It is also clear that it is a matter of judgement as to how to account for non-efficiency considerations. One approach is to limit investigation to aggregate efficiency effects, leaving decision makers to determine the extent to which they are prepared to sacrifice efficiency in pursuit of other objectives. Such an approach may include a qualitative statement of distributional impact.

Bolder approaches attempt to integrate non-efficiency considerations explicitly into analyses by using numerical constraints or differential weights and/or by constructing detailed accounts relating to each objective. Although serious empirical difficulties exist in these approaches, it is not inconceivable that, at least in selected circumstances, generally accepted values for constraints or weights may be generated through a trial-and-error, interactive process between analyst and decision authority, or through expert judgement or public hearings. In Chapter 16 we return to the issue of the use of differential weights in the discussion of social cost–benefit analysis in lesser developed countries where social pricing is in routine use. So far as multi-objective display methods are concerned, there exist several examples of successful implementation, despite early objections to their use.

NOTES

1 The introduction of differential weights may be seen to be another form of shadow pricing, an attempt to measure benefits and costs in terms of 'true' social value.
2 For an explanation of the use of the benefit–cost ratio as the criterion for ranking projects under a budget constraint (assumed here), see Chapter 3.

7

Measurement of benefits and costs: uncertainty and risk

Since the future can never be known with certainty, measurement of benefits and costs raises the issue of uncertainty and risk, especially when benefits and costs have to be estimated over extended periods of time. Uncertainty exists when we know what alternative values benefits and costs could take but the probabilities attaching to these values are not known. Risk exists when possible outcomes are known and, on the basis of either objective evidence from past experience or subjective judgement, the probability distribution of outcomes can be established. For example, in the case of weather impact on park usage, the probability of bad weather for any period can be established from records of the past. In other situations we may be prepared to trust the judgement of experts as to likelihood of outcome.

Before examining alternative ways of dealing with uncertainty and risk, mention should be made of the Arrow–Lind theorem (Arrow & Lind 1970), which suggests that the problem of risk may be ignored in public project appraisal. The argument that governments can adopt a stance of risk-neutrality is that risk is spread across a large number of individuals so that the cost of risk is effectively very low so far as single individuals are concerned. However, the theorem was developed for consideration of financial risk, whereas in CBA there are risks to take into account which may not be capable of being pooled across the whole population. Certain groups of individuals, for example, may be very seriously affected by project failure in so far as project impacts are not diffused entirely through society as a whole. Think of the effect of a nuclear reactor accident on populations in close proximity to the site. Moreover, it is not clear that the Arrow–Lind theorem applies in the case of public goods or bads since risk does not vary in this case with the number of people involved. Pollution, for instance, does not go away because a large number of people rather than a few suffer its effects. Also, risk neutrality is, strictly speaking, justified only if the population is infinite.

In the remainder of this chapter we examine alternative methods of coping with uncertainty and/or risk.

7.1 SIMPLE PROCEDURES

There are three simple and commonly used procedures for taking account of uncertainty or risk. Each involves adoption of a conservative, or play-safe, approach to analysis. The first method involves adding an uncertainty or risk premium to the minimum return required of projects (see Ch. 8), so that a more stringent acceptability crtierion is imposed on projects. The more marginal projects which might have turned out as failures are thus precluded from adoption. This procedure implies that uncertainty or risk increases at a fixed compound rate through time, an assumption which may or may not be appropriate. It also discriminates against projects which yield benefits over the longer term and is, in the last analysis, arbitrary. How do we establish the size of the premium?

A second method for dealing with uncertainty or risk is closely allied to the first and involves truncating the time horizon of the analysis. Since benefits are likely to weigh more heavily than costs in the future, major capital commitments having typically been incurred early in the life of a project, this procedure, like the first, has the effect of imposing on projects a more stringent requirement as to pay-back than would otherwise be imposed. Like the first procedure, this one also discriminates against long-term projects and is arbitrary in design. Furthermore, it may involve treating as identical projects with different uncertainty profiles. If two projects are equally safe up to the arbitrary horizon chosen for the analysis but one of them is more risky thereafter, this difference is ignored under this approach.

The third of the simple approaches to the incorporation of uncertainty or risk into analyses is to estimate benefits and costs on a conservative basis. If in doubt, understate benefits and overstate costs. This procedure provides a safety margin to reflect risk aversion on the part of decision makers although, like the foregoing procedures, it remains arbitrary.

7.2 PROBABILITY ADJUSTMENT FOR RISK

Where there is information about the kind of risks to be anticipated, whether from past evidence or from subjective assessment, probability weights may be attached to estimates of benefits and costs to determine the expected value of the undertaking. The degree of risk may be indicated in terms of the variance or standard deviation of the probability distribution of outcomes. This approach may be implemented using either monetary or cardinal utility measures of benefits and costs.

7.2.1 Monetary measurement of benefits and costs

Table 7.1 Expected value

Possible net outcome $	Probability of outcome	Probability-weighted net outcome $
10	0.10	1.0
20	0.40	8.0
40	0.40	16.0
60	0.10	6.0
	1.00	31.0

The net benefit (or net outcome) and probability data of a single project are shown in Table 7.1. The expected value of this project is $31. The degree of risk involved in the project is calculated as either the variance (V) or standard deviation (σ), where

$$V = \sigma^2 = \sum_{i=1}^{n} P_i (O_i - E)^2 \tag{7.1}$$

where P_i = probability of the ith outcome, O_i = the ith possible outcome, E = expected value of the probability distribution. Thus, continuing the example, $V = \$209$ and $\sigma = \sqrt{V} \approx \$14.5$ (Table 7.2).

Table 7.2 Computation of variance

O_i $	E $	$(O_i - E)$ $	$(O_i - E)^2$ $	P_i $	$P_i(O_i - E)^2$ $
10	31	−21	441	0.10	44.1
20	31	−11	121	0.40	48.4
40	31	9	81	0.40	32.4
60	31	29	841	0.10	84.1
					209.0

Clearly, in situations involving comparison between alternative projects, the project with the highest expected value may not be the one with the lowest degree of risk. In this event, the decision maker has to weigh the trade-off between magnitude of return and risk of achievement.[1] Suppose, for example, two projects as illustrated in Figure 7.1 have equal expected values but different degrees of dispersion. Given risk aversion, project B would be preferred as involving a lower degree of risk.

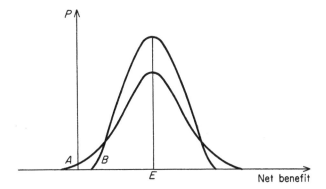

Figure 7.1 Probability distributions.

7.2.2 Cardinal utility measurement of benefits and costs

If greater importance is attached to a loss of a given dollar amount than to a gain of the same amount (because we are risk averse in outlook), irrespective of the probabilities of each outcome, it would be more accurate to measure outcomes in terms of utilities than dollars. The concept of a cardinal utility scale (Friedman & Savage 1948) converts dollar gains and losses into cardinal utility and offers the possibility of using this approach. The scale is illustrated in Figure 7.2 and displays the standard condition of diminishing marginal utility.

For purposes of constructing the scale, a subject is offered a series of choices between hypothetical gambles. Thus, for example, for a gamble yielding $0 or $100 with equal probability, the outcomes are assigned respectively the utility values of 0 and 1 and the subject is asked to specify what dollar sum with certainty corresponds to the expected value of the utility (*EU*) of this gamble, where

$$EU = 0.5 \ (0) + 0.5 \ (1.0) = 0.5.$$

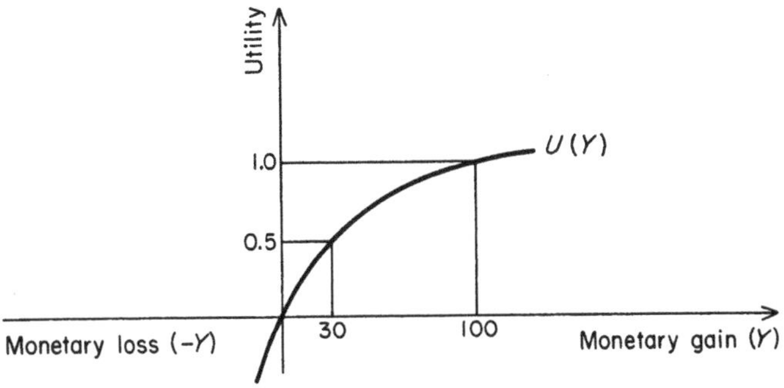

Figure 7.2 Cardinal utility scale.

If this sum is $30 then three points on the utility scale are established:

$$Y_0 : U_0, \; Y_{100} : U_1, \; Y_{30} : U_{0.5}.$$

To find other points on the scale, the subject is asked to identify the probabilities which would make him/her indifferent between other sums to be received with certainty and a gamble. For example, the utility value of $50 is established by computing the expected utility from a gamble involving outcomes of $30 or $100, the utility values of which we already know. Suppose the subject is indifferent between choices when probabilities of 0.7 and 0.3 are attached respectively to the outcomes of this gamble. Then $50 with certainty is equivalent in terms of utility to 0.7 (0.5) + 0.3 (1.0) = 0.65. So Y_{50} corresponds to $U_{0.65}$. By conducting further experiments the utility scale may be traced out in full.

With information on the utility scale, expected values of projects as well as their degree of riskiness may be computed in utility instead of merely dollar terms. Alternatively, risk may be accounted for by deducting from expected monetary value the 'cost of risk bearing'. To see this, suppose in Figure 7.3 that Y_1 and Y_2 are the sole possible outcomes. The points A and B are defined for outcomes Y_1 and Y_2 occurring with certainty, that is $P_{y1} = 1$ and $P_{y2} = 1$ respectively, where, as before, P stands for probability. For $0 < P_y < 1$ for each outcome the broken locus AB is defined showing the expected utility (EU) associated with any gamble (or project) involving outcomes Y_1 and Y_2. Thus if the expected value of the gamble (or project) is $\bar{Y}$, then $EU(\bar{Y})$ is defined where $EU(\bar{Y}) < U(\bar{Y})$. The level of utility derived from the gamble (or project with probability-weighted outcomes) is shown to be equal to that level of utility which would be derived from a lower dollar sum (Y^*) to be received with certainty. The difference $\bar{Y} - Y^*$ is defined as the 'cost of risk bearing'. Thus the criterion for project acceptance that net gain should be positive becomes

$$[\bar{Y} - (\bar{Y} - Y^*)] = Y^* > 0$$

Similarly, certainty-equivalent (or starred) values of benefits and costs can be used in place of expected values in any of the investment decision criteria discussed in Chapter 3.

Practical difficulties with use of the cardinal utility scale are fairly obvious. First, it would seem impracticable to derive scales for a large number of people affected by a project. Secondly, if we rely instead on decision makers' scales it may still be difficult to induce decision makers to engage in time-consuming gambling experiments, especially when these experiments will have to be repeated quite frequently, given that preferences alter over time. Thirdly, it may be wondered whether results will accurately describe how decision makers really would behave in

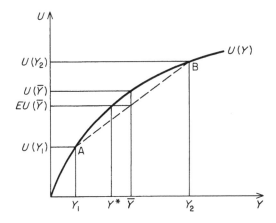

Figure 7.3 Cost of risk bearing.

reality in face of risk. Fourthly, will decision makers' preferences coincide with society's preferences?

7.3 DECISION RULES AND UNCERTAINTY

In situations where there is no confidence in assigning probabilities to outcomes, a selection of decision rules is available for assisting the decision-making effort once estimates of benefits and costs have been made.[2] These rules include maxi-min, minimax regret, maxi-max, Hurwicz and Laplace criteria. An agency's attitude towards uncertainty conditions its choice of decision rule.

We may illustrate use of the different rules with the data in Table 7.3.

Table 7.3 Estimated project outcome

| Projects | Net outcomes | | Row minima |
	a_1	a_2	
A	10	12	10
B	8	13	8
column maxima	10	13	

There are two projects, and a_1 and a_2 represent alternative states of nature, respectively the worse and better conceivable outcomes. The maxi-min decision rule involves maximizing row minima (worst outcomes), a rule which is suitable for the play-safe decision maker, or the one with a pessimistic view of the outcome of nature. In this case project A would be chosen.

Table 7.4   Regret matrix

Projects	a_1	a_2	Maximum regret
A	0	1	1
B	2	0	2

The minimax regret criterion involves derivation of a regret matrix from the information in Table 7.3. This is shown as Table 7.4. The regret matrix displays the difference between the pay-off which could have been achieved with perfect knowledge of the future and the pay-off from any alternative actually chosen. Thus, if we knew with certainty that the state of nature represented by a_1 were to materialize, we should choose project A as having the higher net outcome. In the event that, without perfect knowledge in practice, we do choose project A, we shall have nothing to regret. On the other hand, if we choose project B we shall sacrifice two units of net outcome as compared with what we would have achieved from project A, a regret of 2. Similarly, regret values for a_2 of 1 (for project A) and 0 (for project B) may be generated. The decision rule now requires that we minimize the maximum regret under each possible outcome. Thus we choose project A, the project with the lower maximum regret. This is another rule which suits the conservative or pessimistic decision maker.

The maxi-max rule, by contrast, suits the gambler or optimist. This rule involves maximizing column maxima (see Table 7.3), or 'going for broke'. In this case project B would be chosen. The Hurwicz criterion provides for compromise between pessimism and optimism, allowing the decision maker to specify his degree of pessimism (and hence optimism) in an index (α) of pessimism such that $0 \leqslant \alpha \leqslant 1$ where $\alpha = 0$ represents absolute pessimism about the outcome of nature and $(1 - \alpha) = 1$ represents absolute optimism. The index (α) is then applied as a weight to the worst outcome while the weight $(1 - \alpha)$ is applied to the best outcome, the project yielding the maximum weighted sum of outcomes being chosen.[3] Thus if $\alpha = 0.7$, projects A and B in the example yield weighted pay-offs as follows:

$$\text{project A} \quad 0.3(12) + 0.7(10) = 10.6$$
$$\text{project B} \quad 0.3(13) + 0.7(8) \;\; = \;\; 9.5$$

so that project A would be chosen. The Laplace criterion is a variant on this approach, equal probabilities being attached to expected outcomes on 'the principle of insufficient reason', that is, that assignment of equal probabilities is a reasonable thing to do given that we have no information about the probabilities of outcomes. Applying equal weights of 0.5 to the outcomes for projects A and B yields expected values respectively of 11 and 10.5, so that project A would be chosen. This rule is another which allows for intermediate attitudes towards uncertainty of outlook.

7.4 SENSITIVITY ANALYSIS

As an alternative approach for dealing with risk and uncertainty, or as an approach to be used along with methods outlined previously, sensitivity analysis is always available. Thus results may be presented under a reasonable range of discount rate and time horizon values and/or a range of values for key factors or assumptions in the measurement of benefits and costs. At the same time, the approach may also be used in conjunction with probability adjustments to show expected values on the basis of different probability adjustments, and in conjunction with estimates of different outcomes under the various decision rule methods.

7.5 SUMMARY

A distinction is drawn between (a) a situation of uncertainty in which possible future costs and benefits may be estimated but the probability distribution of outcomes is unknown, and (b) a situation of risk in which probabilities can be established. The Arrow–Lind theorem, widely cited as justifying an attitude of risk-neutrality, is dismissed as misleading in the CBA context.

Simple procedures for dealing with the problems of uncertainty and risk in measuring benefits and costs involve the addition of a premium to the minimum return requirement expected of projects, truncation of the time horizon over which benefits and costs are analysed and estimation of benefits and costs on a conservative basis. While each of these procedures involves arbitrary adjustments, and the first two discriminate against longer-term projects, they are tractable enough to be widely used in practice.

If evidence concerning probabilities of outcome is available, a more satisfactory approach is to compute expected values of benefits and costs along with a measure of dispersion to reflect project riskiness. This approach is readily operational using monetary evaluation of benefits and costs, although in principle it may also be implemented using utility measures of project outcome (drawing on the concept of the cardinal utility scale). An alternative approach, also drawing on the utility scale, is to compute the 'cost of risk bearing' and to express costs and benefits in terms of certainty-equivalent rather than expected values. The difficulties in constructing the utility scale, however, give these latter approaches more conceptual than practical value.

In situations of uncertainty, various decision rules drawn from game theory are available for use by decision makers presented with estimates of benefit and cost under different outcome scenarios. Choice of rule depends on decision makers' attitudes towards uncertainty; whether they are gamblers or not; and whether they are optimists or not.

NOTES

1 A mechanical rule combining in a single index both expected value and dispersion measures (e.g. the coefficient of variation) may be employed to guide decisions.
2 The material in this section follows closely the development in Pearce 1971, 1983, Pearce & Nash 1981.
3 In cases involving more than two possible outcomes, the best and worst outcomes alone are weighted.

8

Measurement of benefits and costs: minimum return requirement

In order to make annual flows of benefit and cost commensurable for purposes of aggregation over time, it is necessary, as pointed out in Chapter 3, to take account of the time value of money. Money received or paid early is worth more than money received or paid later. This is because money can be used to earn interest. Thus earlier receipts grow at compound interest to a greater sum at a given time horizon than later receipts. Similarly, earlier payments cannot be invested to earn as much interest as delayed payments. Delayed payments are therefore effectively lower in cost if accrued interest is deducted from them. In other words, a given sum of money has different value depending on when it is received or paid; and in order to aggregate the stream of benefits generated or the stream of costs incurred over the life of a project, annual flows must be expressed in equivalent values.

The usual procedure is to express benefits and costs in terms of their present value: their monetary equivalent if received or paid in the present. It is possible, however, to make annual flows comparable in terms of any point in time; and, again as indicated in Chapter 3, terminal rather than present value is sometimes used in CBA. Benefits and costs are in this case expressed in terms of their monetary equivalent in the last year of a project's life.

In order to compute present (or terminal) value, an appropriate discount (or interest) rate is required. The rate to be used is the minimum rate of return (MRR) that a public investment is required to earn if it is to be worthwhile.[1] Unless the project is capable of earning this rate of return (interest) or, the same thing, being discounted at this discount rate, the resources it uses would be better employed in an alternative use. In principle, there is little difficulty in identifying the MRR as the opportunity cost of the funds used on the project. In practice, difficulties arise from two sources. First, there are two underlying concepts with a claim to providing a basis for establishment of the MRR, and typically each generates a different value for MRR. Secondly, the rates suggested by the underlying concepts have to be measured. As one commentary puts it, 'whatever one does, one is trying to unscramble an omelette, and no one has yet invented a uniquely superior way of doing this' (Prest & Turvey 1965, p. 700).

In this chapter we provide a summary of alternative approaches to these problems, dealing in turn with the conceptual basis of the MRR, methods for reconciling use of the different rates suggested, approaches to the problem of measurement and, finally, some procedures used in practice.

8.1 UNDERLYING CONCEPTS

In so far as funds for a public sector project are derived ultimately from the private sector, they may be seen as displacing private sector investment or current consumption opportunities.[2] Determination of the MRR on public sector proposals should, therefore, be based on the value of the private sector opportunity forgone. To the extent that private investment is displaced, the social opportunity cost rate (SOC) is relevant, and to the extent that private consumption is displaced, the social time preference rate (STP) is relevant. The SOC is defined as the marginal pre-tax rate of return in the private sector, the rate of return which the funds would have yielded for society, abstracting from externalities, had they not been commandeered by the public sector. The STP is defined as the rate of return required by society to induce it to sacrifice present consumption for the promise of future consumption as generated through investment. Since the STP varies with preferences for present relative to future consumption, it may be expected to depend on, inter alia, the general level of prosperity, expectations regarding the growth of prosperity, the age distribution of the population (the older the population, the less important the future) and altruism so far as concern for future generations goes.

Taking a simple two-period model, the concepts of SOC and STP may be illustrated as in Figure 8.1. The fact that the two rates are typically not equal in practice is also illustrated. PP' represents the transformation function between current and future consumption as faced by society. It is drawn as a straight line for simplicity, although in practice it is more likely to be concave to the origin, reflecting the principle of increasing opportunity cost. Society may enjoy OP' units of current consumption and no future consumption, or OP units of consumption next year if it sacrifices current consumption for investment, or the combinations of C_t and C_{t+1} as indicated by the frontier PP'. Moving left on the abscissa indicates the amount of investment (I_t) undertaken in period t. II' is the social indifference curve showing the rate at which society is willing to sacrifice current for future consumption.

It can be demonstrated that SOC = slope of PP' − 1. It can also be

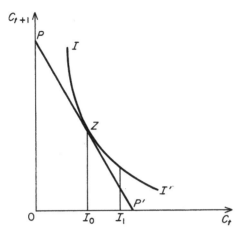

Figure 8.1 Rates of SOC and STP.

demonstrated that STP = slope of II' $-$ 1.[3] It is clear in Figure 8.1 that $SOC = STP$ solely at the welfare maximizing point Z when I_0P' units of current consumption are sacrificed to investment. In practice, society may typically be undertaking less than the optimal amount of investment, say I_1P'. Reasons for underinvestment may be related to savings inadequacy, imperfect capital markets which cause less borrowing than society would wish to incur, or monopoly influences which constitute an impediment to the influx of capital to some sectors. Another reason for discrepancy between SOC and STP rates is that SOC is defined before corporate taxes are deducted from returns which accrue to individuals, while STP is defined after such taxes are removed. In any case, at the level of investment I_1P', the slope of PP' is greater than the slope of II' so that $SOC > STP$. That SOC is typically greater than STP constitutes a fundamental difficulty in the choice of the social discount rate or MRR.

8.2 RESOLUTION OF PROBLEM OF DIFFERENT RATES

Two methods are currently proposed for combining SOC and STP in the determination of the social rate of discount.[4] The first method may be termed the weighted average cost of funds approach. This involves weighting the costs of funds that displace respectively private investment and private consumption by the proportions of total funds from each source. The cost of funds that displace investment is the SOC while the cost of funds that displace consumption is the STP.[5]

Aside from the problems of measuring SOC and STP (which we come to in Section 8.3), there exists the difficulty of identifying the proportions of funds which displace investment and consumption. One suggestion in this regard is to use as the basis of establishing proportions the division of public receipts between personal taxes (which affect mainly consumption) and borrowing and other taxes (taken to affect mainly investment). This is obviously a rather crude approach. Another alternative is to employ sensitivity analysis, undertaking calculations according to a reasonable range of values.

The second method for combining *SOC* and *STP* may be termed the shadow price approach. This involves shadow pricing capital outlays on the project in terms of equivalent forgone consumption and discounting inter-temporal effects by *STP*, the appropriate rate given that all project flows are in the form of equivalent consumption. Thus, following Eckstein (1961a), if *SOC* = 6% and forgone private investment returns (forgone consumption) are assumed in perpetuity, and if *STP* = 3%, then the present value of the forgone consumption from a capital outlay of one dollar on the project = $1(0.06/0.03) = $2, or the shadow cost adjustment factor = *SOC/STP*. Net annual benefit (B_t) and adjusted capital cost flows (K) are then discounted at 3%. Thus, the net present value (*NPV*) of a project is defined as:

$$NPV = \sum_t \frac{B_t - (SOC/STP)K_t}{(1 + STP)^t} \qquad (8.1)$$

If it is considered appropriate to distinguish between displacement of private investment and private consumption, the shadow cost adjustment factor (Marglin 1963b) becomes:

$$\theta \left[\frac{SOC}{STP} \right] + (1 - \theta) \qquad (8.2)$$

where θ = the proportion of expenditure which displaces private investment. The assumption is made here that forgone investment returns occur in perpetuity. However, the approach may be modified to include a finite time horizon on forgone returns (Marglin 1963b). It is also possible to incorporate into the shadow cost adjustment factor recognition of the possibility that some proportion of the annual return from investment in both private and public sectors is reinvested (Marglin 1963b). If foreign capital provides part of the funding for the project, the forgone consumption associated with providing a return on it may be dealt with in the manner of forgone consumption on displaced private investment.

It may be noted that the weighted average and shadow price approaches yield identical investment criteria in the case of perpetuities in both private and public sectors. However, when public sector projects have finite lives, the shadow price approach is the more stringent (Sjaastad & Wisecarver 1977). In the case of perpetuities, leaving out foreign funding for simplicity, the weighted average discount rate (*W*) is:

$$W = \theta \, SOC + (1 - \theta) \, STP \qquad (8.3)$$

Under the shadow price approach, the adjustment factor (α) is given as:

$$\alpha = \theta \left[\frac{SOC}{STP} \right] + (1 - \theta) \qquad (8.4)$$

so that assuming capital cost to be incurred entirely in the initial period, the net present value (*NPV*) criterion is:

$$B_t/STP \geqslant \alpha K_t \qquad \text{or} \qquad B_t/STP\alpha \geqslant K_t \qquad (8.5)$$

The implicit discount rate is, therefore:

$$STP\alpha = \theta \, SOC + (1 - \theta) \, STP = W \qquad (8.6)$$

indicating that the two approaches are identical.

When public sector projects have finite lives, however, the implicit discount rate using the shadow price approach becomes ($\alpha + W$), a more stringent discounting requirement than the weighted average rate of W. To illustrate, take the polar case of a public investment project which yields the whole of its benefits in the period after the investment is made. The *NPV* criterion then becomes:

$$B_t/(1 + STP) \geqslant \alpha K_t \qquad \text{or} \qquad B_t/\alpha(1 + STP) \geqslant K_t \qquad (8.7)$$

giving an implicit discount rate of $\alpha(1 + STP) = (\alpha + W)$.

Doubtless, because the weighted average approach seems to be the more easily comprehensible, it is the weighted average discount rate which is commonly used in practice.

8.3 MEASUREMENT OF *SOC* AND *STP*

The problem of measuring the rates which form the basis of the determination of the social discount rate or *MRR* has up to this point been avoided. We now address this issue, looking first at *SOC* and then at *STP*.

8.3.1 SOC *measurement*

It is clear in principle that *SOC* represents the pre-tax marginal rate of return on investment in the private sector. In principle, too, this rate should be adjusted for externalities since we are concerned with the forgone social rather than private rate of return. For the same reason, the pre-tax rather than post-tax rate of return is relevant.[6]

Approaches to the measurement of the required rate make a number of concessions to data constraints. There are two general approaches. The first involves use of an estimate of the private sector cost of capital as representing the corporate discount rate and, to the extent that investment is undertaken up to the point at which the rate of return equals the cost of capital, also represents the rate of return at the margin. The post-tax cost of capital as estimated in the corporate sector is grossed up for corporation, capital gains and dividend withholding taxes. The estimation of the cost of capital is based on the weighted average cost of debt and equity funds where the cost of the latter is based on estimates of the average rate of return achieved on the stock market (Merrett & Sykes 1966). The second approach is to estimate the average (rather than marginal) rate of return in the private sector using the ratio of pre-tax profit to the value of capital stock (Jenkins 1973). It is clear that all manner of approximations are involved in both approaches.

8.3.2 STP *measurement*

Measurement of *STP* is based on either after-tax returns to private saving or the marginal utility of consumption. So far as the first approach is concerned, it is argued that after-tax returns reflect people's time preferences in so far as they are willing to give up current consumption in return for the net return offered. The long-term government borrowing rate is sometimes suggested as the appropriate rate on the grounds that it is a riskless rate and (a) for public projects risk is negligible (Arrow & Lind 1970), or (b) risk should be taken into account in ways other than through the discount rate (see Ch. 7 above). If it is preferred to incorporate some recognition of risk into the time preference rate, a weighted average of after-tax public and private sector bond rates and the average after-tax return on equities, or some subset of these values, may be used.

This said, it is to be recognized that arguments have been advanced for using rates either lower or higher than observed rates. So far as lower rates are concerned, there are two arguments (Marglin 1963a). The first is that individuals suffer from myopic time preferences or a 'faulty telescopic faculty' (Pigou 1932), impatience compelling them to seek higher returns when they make private investment decisions than may be optimal in light of the interests of future generations. An implication of this argument is that a zero, or even negative, discount rate may be appropriate for inter-generational proposals. Otherwise, proposals which impose a huge cost on future generations (for example, loss of non-renewable or unique natural resources) may be justified in terms of benefits for the current generation, given that the loss will occur so far in the future that its present value is trivial. Similarly, proposals yielding a huge gain for following generations at relatively little cost to the current generation could be rejected.

The second argument for a lower than observed rate is based on the 'isolation paradox' (Sen 1967). The idea is that collective decision making may reveal a lower rate of time preference than is revealed through the market in private investment decisions. According to the paradox, individuals may not be willing in isolation to undertake investment for the benefit of future generations, yet they may subscribe to a collective agreement to do so. One explanation for this is that if an individual's utility level depends in part on consumption levels of others relative to his/her own, the individual could be reluctant to sacrifice current consumption in order to fund investment spending unless others are prepared to do the same. Another way of explaining the 'isolation paradox' is to visualize investment for the benefit of posterity as a public good that is psychically consumed by all members of the community and therefore subject to the 'free rider' problem. Either way, the social rate of time preference would be lower than market-revealed private rates.

Arguments for using, by contrast, a rate which is higher than market-revealed rates have also been made. First, market rates fail to capture a representative cross section of private time preference rates to the extent that it is the better-off members of society who are active in financial markets and they will tend to have lower rates of time preference than the less well off. Secondly, the lower the rate employed, the more perverse are its implications in terms of requiring the sacrifice of present for future consumption. If future generations are likely to be better off than the current generation, the lower the rate employed the more its use involves robbing the poor (today's generation) to give to the rich (tomorrow's generation). When all the arguments are considered, then, it is not clear what precise rate would emerge from a starting basis of market rates of interest. Hence, judgement is called for on the part of analysts.

An alternative approach to the measurement of *STP* is to estimate it from an expression for the marginal utility of consumption (Scott 1977, Pearce & Nash 1981). The marginal utility of current consumption (MU_{ct}) is expressed as a function of current consumption per capita (C_t):

$$MU_{ct} = aC_t^b \tag{8.8}$$

where a and b are constants, the latter the elasticity of the marginal utility of consumption function. Given that

$$STP = (MU_{ct}/MU_{ct+1}) - 1$$

(see note 3 below), it follows that:

$$1 + STP = aC_t^b/aC_{t+1}^b{}^b = (C_{t+1}/C_t)^{-b} \qquad (8.9)$$

Let $(C_{t+1}/C_t) = 1 + c$ where c = the proportionate rate of growth of consumption per capita. Thus:

$$1 + STP = (1 + c)^{-b} \qquad (8.10)$$

or

$$STP = (1 + c)^{-b} - 1 \qquad (8.11)$$

Using an assumed value for b (typically -1 or -2), it is a simple matter to estimate STP from data on the proportionate rate of growth of consumption per capita. However, approximation is again involved, particularly regarding the value for b.

8.4 SENSITIVITY ANALYSIS

Given all the difficulties, both conceptual and empirical, in estimating MRR, the usual procedure in practice is to adopt sensitivity analysis. Calculations are performed over what appears to be a reasonable range of discount rates, having in mind the considerations raised in this chapter.

Good examples of the use of sensitivity analysis lie in the practices recommended for analysis of federal and provincial government projects in Canada. A 5–15% range of real discount rates is recommended for federal government proposals (Government of Canada 1976) with calculations also to be done at a benchmark median value of 10%. This recommendation is derived from an estimate of the social discount rate of approximately 9.5% as based on a weighted average of the pre-tax rate of return on displaced private investment, the post-tax rate of return on saving (forgone current consumption) and the rate of return required to induce incremental foreign funding (Jenkins 1973).[7] The rate of return on displaced investment is estimated as the average sectoral ratio of pre-tax profit to capital stock (at replacement cost). In the Canadian province of British Columbia a range of real rates of 8–12% is recommended with a benchmark rate of 10%, again derived from Jenkins' work (Province of British Columbia 1977).[8]

8.5 SUMMARY

It is apparent that numerous difficulties attach in practice to the precise determination of the social rate of discount. The conceptual underpinnings of the appropriate rate, however, are clear and a number of suggestions exist as to procedures to follow in order to establish a reasonable framework for choice of the rate.

The rate that is sought is based on the *SOC* and *STP* rates for funds invested in projects. Unfortunately, these two rates in practice provide different values. However, the discrepancy in rates may be addressed by either taking a weighted average of the two (with weights being proportions of funds displacing respectively private investment and consumption spending) or by shadow pricing capital costs in terms of the *SOC* rate and discounting net annual benefits at the *STP* rate. If foreign capital provides a source of part of the funding for a project, it is necessary to include it in a weighted averaging of costs or in a shadow cost adjustment factor reflecting forgone consumption, depending on the method in use. Its cost is its net-of-tax supply price.

Empirical measurement of *SOC* and *STP* rates presents further difficulties. The *SOC* rate may be measured, either by estimation of the private sector pre-tax cost of capital (which is equated to the private sector rate of return at the margin), or by estimation of the pre-tax average ratio of profit to capital cost. The *STP* rate may be measured in terms of after-tax market rates of return on private savings, though arguments exist for adjusting these both upwards and downwards. An alternative approach is to estimate *STP* from an expression for the marginal utility of consumption. The safest procedure for choosing *MRR* is to employ sensitivity analysis over a range of rates based on considerations outlined above. Examples from actual experience are provided to illustrate this approach.

NOTES

1 If annual benefits and costs are measured in real (constant dollar) rather than nominal terms, then the *MRR* is also to be expressed in real terms.
2 For simplicity, we ignore here the possibility that funds might be drawn from foreign sources as well as from alternative domestic uses. Foreign funding may be readily taken into account in practice, as the subsequent discussion indicates.
3 Since the slope of $PP' = OP/OP' = C_{t+1}/I_t$ = rate of return on investment $+ 1 = SOC + 1$, SOC = slope of $PP' - 1$. Since the slope of $II' = MU_{ct}/MU_{ct+1} = STP + 1$, STP = slope of $II' - 1$.
4 In some early work (e.g. Baumol 1968) it was advocated that the lost private opportunity with the highest cost is the only relevant measure of the opportunity cost of funds used for public projects, so that given $SOC > STP$, *STP* may be disregarded. It is now recognized, however, that the actual rather than the highest opportunity forgone is the appropriate measure of opportunity cost in a second-best world (Feldstein 1972).
5 The weighted averaging procedure may also take account of foreign funding if relevant (e.g. Jenkins 1973), the cost of which is the net of tax supply price of foreign capital, that is, the return necessary to attract it.
6 In so far as benefits and costs are measured in real terms, the discount rate should also be so measured (see note 1 above).
7 For readers interested in following up a discussion on the validity of this approach, see Sumner (1980) and Burgess (1981). Sumner suggests that the median value of 10% may be too high, given that (a) Jenkins' results suggest

9.5%; (b) the approach estimates the average rather than the marginal rate of return; (c) monopoly barriers in some sectors sustain the rate of return above the level necessary to induce new investment spending; and (d) the tax adjustments made by Jenkins in arriving at pre-tax profit overstate gross return. Burgess also suggests that the recommended rate may be too high due to (a) understatement of the contribution of savings and foreign funding to the financing of public investment; and (b) overstatement of the opportunity costs of all types of funding.

8 It may be noted that the benchmark rate of 10% corresponds exactly to a rate in use by the US government and the test discount rate used for several years for public projects in the UK.

PART II

Applications of cost–benefit analysis in urban and regional planning

9

Residential urban
renewal

We begin our review of the prospects and pitfalls of applying cost–benefit methods to issues of urban and regional planning with an examination of the use of CBA to evaluate schemes for improving the quality of urban housing. These schemes involve replacement of a deteriorated housing stock with a new one, through slum clearance or rehabilitation of the old stock. In this chapter we assess the traditional framework for analysing the classic case of residential slum clearance (Rothenberg 1965, 1967) before looking at additional models for evaluating housing replacement or rehabilitation schemes.

9.1   ROTHENBERG MODEL

Adopting the viewpoint of the metropolitan economy, Rothenberg listed the aggregate efficiency effects of redevelopment shown in Table 9.1. Improved productivity or value of land at the renewal site represents increased producer surplus (economic rent) for landowners. Ignoring consumers' surplus, this, in turn, reflects what people are willing to pay for use of the improved site.[1] In addition, there are external benefits in the form of increased value of neighbourhood land and properties as a result of area upgrading, and reduced social costs of health, fire and crime hazard. Resource costs comprise site acquisition cost (less the initial value of the land since this is not lost to society through the renewal project) together with redevelopment expenditures on the site.

On the basis of this framework, Rothenberg undertook an *ex post* evaluation of three urban renewal projects in Chicago, while Messner (1968) evaluated various projects in Indianapolis. In both studies only site

Table 9.1   Benefits and costs of residential urban renewal

Benefits	Costs
increased site productivity neighbourhood spillovers reduced social costs	project resource costs

productivity benefits were estimated, using the observed change in site value with corrections for the following factors:

(a) population, income or other exogenous changes during the period of redevelopment as these influenced land values;
(b) capitalized property taxes as these would be deducted from observed market value; and
(c) the estimated effect of changed locational advantage as a result of a project.

In the latter case, an urban renewal project may be expected to enhance the value of site land partly because the redeveloped site attracts economic activity which would otherwise have occurred elsewhere in the metropolitan area. Any land price increase associated in this way with a change in intra-metropolitan locational advantage, or indeed with any other pecuniary effect which is netted out from the aggregate viewpoint, has to be removed from the estimate of benefit, unless project neutrality in this respect is a reasonable assumption.

Table 9.2 Indianapolis: redevelopment projects ($000 at present value)

	Projects		
	A	B	C
resource costs	1590	162	817
project benefits			
site productivity	239	27	183
spillovers	m	m	m
reduced social costs	m and i	m and i	m and i
costs not offset by measured benefits	1351	135	634

m = measurable but not measured; i = intangible.
Source: Messner 1968, p. 156.

For illustrative purposes, certain of Messner's findings are shown in Table 9.2. Estimates of the increased value of neighbourhood land and properties, and of reduced social costs, were not attempted. Given this partial accounting on the benefit side, it was not possible to present a clear-cut ranking of alternative schemes. The bottom line of the analysis identified merely costs not offset by site land benefits, and it was left to decision makers to judge whether the likely magnitude of the intangible and otherwise unmeasured benefits at each site would be sufficient to make up the shortfall. It may be noted how relatively insignificant in the analyses were land value effects, suggesting the importance of reduced social costs or other factors in justifying renewal projects.

Mao (1966) extended the application of the same basic framework by attempting an estimate of the benefits of neighbourhood spillovers and of reduced social costs associated with an urban renewal project in East Stockton, California. Rough measurement of the increase in value of neighbourhood properties was achieved by interviewing real estate experts about properties which they knew. For purposes of measuring reduced social costs it was assumed that the annual cost per resident of fire, police and health protection after renewal fell to the average for the city of Stockton as a whole. This procedure clearly attempted to capture only the financial savings accruing to the city from reduced social costs, abstracting from benefits of reduced hazard to individuals involved. The ideal procedure would have gauged willingness-to-pay for reduced hazard using, say, the contingent valuation method, or avoided property loss data and the valuation of reduced risk to life and limb (see Ch. 10). It may be noted, in addition, that Mao's method failed to take account of the possibility of offsetting increases in social costs elsewhere in the city as displaced former residents moved away from the site and the 'filtering down' process got under way. Moreover, it is advisable when dealing with relatively intangible project impacts to supplement tentative money measurements with indicators of non-monetary effects; for example, changes in the incidence of crime, fire and illness.

While the Rothenberg framework provides a conveniently simple picture of economic benefits and costs of residential urban renewal, it is clear that it presents assorted difficulties in estimation and is less than comprehensive. Aside from the problems of measuring neighbourhood spillovers and reduced social costs, a number of complications arise in accounting for non-renewal influences (e.g. population and income changes) on changed land prices, and for pecuniary effects such as changes in locational advantage within the city. Furthermore, land values need to be shadow priced if land is not bought and sold for the scheme on the basis of highest bids. Land is often acquired by the local authority at less than market value through powers of compulsory purchase and is not then sold for private development; instead, the local authority itself, as would be the typical case in Britain, undertakes the development. One way around this problem is to prepare alternative estimates of resident benefits (Section 9.2 below) which capture site user effects not adequately reflected in differential land values. Finally, estimates of value are likely to be more tenuous in *ex ante* than in *ex post* studies, although sensitivity analysis and other procedures discussed in Chapter 7 for dealing with uncertainty are available for use.[2]

As to the scope of the analysis, a comprehensive approach would compare alternative methods of achieving the objective of urban renewal rather than merely comparing specific renewal projects with the status quo. It may be preferable, for instance, to alter zoning regulations or to renovate the existing housing stock instead of engaging in wholesale redevelopment. Thus Case (1968) adapted the Rothenberg framework to the analysis of building code enforcement as an alternative means of urban renewal. For an illustration of the comparison of rehabilitation and redevelopment as measures to improve housing conditions, see Section 9.3 below. It may also be noted that the Rothenberg framework is not concerned with the distribution of benefits and costs as between different groups affected, although in principle these may be identified. The viewpoint of the analysis is also confined to the metropolitan area, whereas a different community viewpoint (e.g. the province or state, region, nation) may be relevant, depending on the purpose of the project under analysis and the sources of its funding. Again, however, the framework may be applied in principle to a broader – or for that matter, narrower – perspective. The same observation applies in respect of types of urban renewal project other than the conversion of an old residential area into a new one (see Ch. 12). While the scope of the Rothenberg analysis may not, therefore, be as broad as would be ideal, that is not to say that it cannot be broadened in practice.

So far as the definition of benefits goes, the approach provides limited treatment of external impacts. In practice even purely residential renewal schemes are likely to generate certain non-housing effects not included in the framework of analysis: for example, the possible relief of traffic congestion; recreational user benefits where part of the redeveloped site is dedicated to park or other recreational usage; income created in the city as a result of investment spending which would not have been attracted to the city in the absence of the project. Measurement of these items is discussed in other chapters, and there is no reason in principle why the approaches outlined in these separate discussions should not be employed where appropriate to facilitate more complete analyses of residential urban renewal projects. In terms of housing effects themselves, the approach also omits dislocation costs incurred by displaced residents of original properties.

In the following section we discuss the measurement of these dislocation costs for original residents together with the measurement of benefits for residents of the new housing. Procedures discussed for measurement of the latter may be seen as alternatives to the measurement of site productivity benefits as indicators of on-site welfare gain.

9.2   NET RESIDENT BENEFITS

When a slum removal project occurs, residents suffer dislocation costs. These include intangible, psychic costs associated with being uprooted, along with the monetary costs of moving and perhaps of having to pay higher rents and/or daily travelling expenses elsewhere. After renewal, there are presumed to be welfare benefits for new residents who are able to enjoy better housing conditions for the price than they would find elsewhere. Looking back, of course, on the experience of building cheap, often soulless high-rise blocks in place of traditional neighbourhoods, it is apparent that these presumed benefits may not always have been positive. Our concern in this section is to examine the possibility of measuring directly the net effect on the welfare of both new and displaced residents, a consideration not incorporated satisfactorily into the Rothenberg model.

9.2.1 New residents: welfare gain

Let us deal first with benefits for residents of the new housing, leaving dislocation costs for displaced residents (who may or may not be the same people) for separate discussion. Assuming, due to public ownership or subsidies to private ownership, that the scheme creates housing that is less expensive per unit of quality than would otherwise be available in the unsubsidized market, net benefits may be measured in terms of consumers' surplus (Ch. 4). Methods employed to estimate resident surplus have involved estimation of:

(a) the ordinary Marshallian measure of surplus, the area beneath the demand curve and between price lines (e.g. Sumka & Stegman 1978, Kraft & Kraft 1979); and

(b) the Hicksian measures of equivalent variation (e.g. DeSalvo 1971, 1975, Murray 1975, Walden 1981) and compensating surplus (Flowerdew & Rodriguez 1978).

Looking at the Marshallian measure of surplus, Figure 9.1 shows a single household's demand curve for housing, the quantity of housing service comprising size, appearance, location and other relevant physical features that constitute the quality of a housing unit. P_m is the market price facing the household in the absence of the renewal scheme; P_s is the price per unit of housing service in the scheme. Resident net benefit (B) is the sum of areas (a) and (b), respectively the money saving on the quantity of housing service which would have been purchased in the absence of the project and the surplus on additional units of service enjoyed as a result of the project.

Assuming a unitary price elasticity so that the demand curve is a rectangular hyperbola:

$$B = P_m Q_m - P_s Q_s + P_m Q_m [l_n(P_m Q_s) - l_n(P_m Q_m)] \qquad (9.1)$$

where l_n indicates logarithms. For ease of computation, an alternative assumption may be that the demand curve is linear. Estimates of benefit for individual households are then aggregated over households affected.

The market value equivalent of subsidized housing $(P_m Q_s)$ may be estimated from either the judgement of housing authority personnel or a

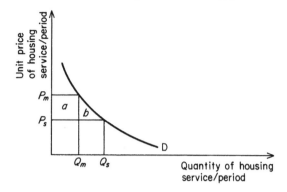

Figure 9.1   Household demand for housing services.

hedonic regression explaining private market rents on the basis of housing characteristics. The counterfactual value of expenditure for housing in the absence of subsidized units $(P_m Q_m)$ may be estimated from regression analysis using residents' socio-economic characteristics as determinants of private market housing consumption. Alternatively, information regarding previous private market expenditure as a proxy for counterfactual expenditure may be derived by survey. The method is clearly subject to some imprecision in measuring these unobservable values.

Moreover, as developed above, the method assumes that the household has a free choice as to quality of housing occupied in the renewal scheme. Clearly, it may be more realistic to recognize that type of accommodation is often a given; it is offered on a take-it-or-leave-it basis so that the household is required to purchase a lesser or greater amount of housing

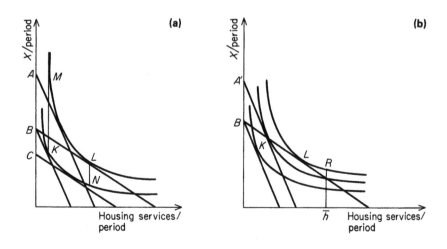

Figure 9.2   Resident consumer surplus.

service than would really be preferred at the prevailing price. In this event, the household is forced off its true demand curve and the empirical measure of surplus based on actual rather than preferred units of consumption becomes only an approximation of true surplus. Also, use of the Marshallian demand curve assumes a zero income effect, not that this is necessarily a serious flaw (Ch. 4).

Measurement of resident benefits in terms of the equivalent variation involves estimation of the compensation necessary to make residents as well off without the urban renewal scheme as with it. In Figure 9.2a a composite commodity (X), or money available for expenditure on all commodities other than housing, is arrayed on the vertical axis. In the absence of the renewal scheme, the household would choose the combination of the composite commodity and housing represented by point K. After renewal, the combination L is chosen, assuming that renewal makes available a unit of housing service at a lower price than would be available elsewhere and that the household has a free choice as to the type of accommodation acquired. AB is the equivalent variation measure of consumer surplus. For the record, BC is the compensating variation, LN the compensating surplus and MK the equivalent surplus (see Ch. 4).

In Figure 9.2b, the case of the take-it-or-leave-it offer is illustrated. The household either accepts a dwelling which embodies services at the level $(\bar{h})$ or does not participate in the scheme at all. Participation puts the household at position R yielding a lower level of welfare than would have been attained (at L) with freedom of choice as to housing quality. It may be noted that the level of quality $(\bar{h})$ could be equal to the level that would have been freely chosen, or lower than that level. Unless $(\bar{h})$ is equal to the preferred level, the equivalent variation is smaller than the measure of equivalent variation in Figure 9.2a. In the illustrated case the equivalent variation is $A'B$.

In a pioneering study of resident benefits of a New York City subsidized housing programme, DeSalvo (1971, 1975) operationalized the equivalent variation approach by assuming for convenience a Cobb–Douglas utility function in which the exponents sum to unity and unitary price and income elasticities are implied. Thus:

$$U = h^\beta X^{1-\beta} \tag{9.2}$$

where U = household utility, β = rent/income ratio in the absence of the programme, h = flow of housing services, and X = composite commodity. While empirical evidence suggested that the implied elasticities were not unreasonable, more complex utility functions could be employed if preferred (see, for example, Murray 1975, Olsen & Barton 1983, DeBorger 1985). By maximizing the utility function subject to an income (Y) constraint, substituting the resultant demand equations into the utility function and solving for Y, the equivalent variation (or, if required, the compensating variation) can be computed from the rental price of a project housing unit, the market price equivalent of the unit, resident income and the rent/income ratio (β).[3]

As with the Marshallian method, it is necessary to estimate unobservable values (equivalent market price and the rent/income ratio). For the former, DeSalvo used expert judgement and, for the latter, econometric estimation of a rent/income function including family characteristics as independent variables. His results appear in Table 9.3. An alternative way of estimating the rent/income ratio is to determine *previous* levels of unsubsidized rent by survey (Walden 1981).

Table 9.3 Annual benefits and costs of a housing project

	All households $m 1968	
Benefits		
resident consumer surplus	25.6	
project rent	84.0	
gross resident benefits		109.6
Costs		
project rent	84.0	
public subsidy	46.9	
total resource costs		130.9
Minimum required non-resident benefits		21.3

Source: DeSalvo 1975, p. 801.

Of course, the foregoing methods of resident surplus estimation assume that the observed rental or purchase price of a renewal dwelling is lower than the price of comparable accommodation elsewhere, a condition which may not always hold. Housing in the renewal scheme may not be subsidized and a household may choose to live in such housing, not because of lower housing price, but because of perceived net advantages when all relevant considerations (price, commuting and other associated expenses, general amenity, and so on) are taken into account. This situation may still be interpreted in the context of the models discussed, with price defined to include ancillary living expenses and housing service recognized to include all intangible as well as tangible features of housing units. However, if it is necessary to analyse a project on this basis, the measurement of relevant prices and quantities becomes that much more difficult using the empirical methods described above.

An alternative method for estimating Hicksian measures of resident surplus, and one which is less demanding in terms of data requirements, is contingent valuation. Flowerdew & Rodriguez (1978) used this method to estimate the compensating surplus for residents of a scheme which replaced a cluster of Victorian residential properties with council (or public) housing. A direct interview procedure was employed following an initial questionnaire survey designed to determine whether the move to the site had been a good or a bad experience. If the latter, respondents were asked whether the move would have been regarded as a good thing (and vice versa for a good move) if the rent/period had been lower (higher) by any of the following prices: 25p, 50p, £1, £2 and £4. Starting prices were chosen randomly to avoid any response bias which might have attached to the order in which alternative price possibilities were suggested. If the answer was positive, respondents were asked the same question with respect to the next lower (higher) price, and so on until a negative response was given. Estimates of consumer surplus gained were placed at the mid-point between accepted and rejected prices.

The possibility of strategic bias remains. Residents, for example, may have understated their estimates of surplus fearing that rents might be adjusted to reflect them; or they might have overstated them, knowing that they would not have to pay. Even so, the approach illustrates a method of usefully supplementing the traditional Rothenberg approach with a measure of net resident benefits. It is to be noted that the method may be applied to the case of residents of neighbourhood as well as site properties, and that the method also measures residents' willingness-to-pay for such benefits as reduced hazards of crime, fire and disease and savings (if any) in miscellaneous living expenses as a result of moving to the renovated site.

9.2.2 Displaced residents: welfare loss

As explained earlier, it may be necessary to estimate dislocation costs for original residents displaced from the renewal site. One method of estimation is to focus on the monetary costs involved in displacement: the costs of removal, extra accommodation costs if applicable and increased costs of commuting and shopping if applicable, each of which is relatively easy to estimate through questionnaire surveys. However, this approach provides merely a lower bound estimate of dislocation costs, the intangible costs of disruption (the break-up of social relationships, separation from a known environment) being omitted. In order to include these costs, it is necessary to measure the loss of consumer surplus.

If displaced residents end up paying higher rents and incurring other additional expenses, it is possible to measure their welfare loss on the basis of Marshallian or Hicksian measures of surplus. In principle, methods analogous to those described for estimating resident benefits may be used, although to date the contingent valuation method alone appears to have been tried.

A well-known example of the use of contingent valuation in the displacement context came from the Roskill Commission (Commission on the Third London Airport 1971) in an attempt to gauge the loss of welfare to residents who would be displaced by the development of a third London airport. Unfortunately from the point of view of promoting the method, the attempt was marred by simplicity and has been heavily criticized (e.g. Mishan 1970). When asked hypothetically what sum they would require by way of compensation, for example, respondents were given no indication of how far they would have to move in the event of being displaced so that they could not adequately assess all the cost implications of moving. Moreover, answers considered by the research team to be 'too high' were arbitrarily lowered. Even without elementary shortcomings of this kind, the method remains open to all the usual difficulties of contingent market evaluation as these have been outlined in Chapter 5. A careful example of the use of the method is provided by Flowerdew & Rodriguez (1978), who used it to estimate welfare effects of moves from a renewal site, as well as of moves to the site as described earlier.

9.3   COST-EFFECTIVENESS ANALYSIS

A separate strand of analysis in the field of housing and urban renewal falls into the category of cost-effectiveness analysis, concentrating on the minimization of cost subject to the provision of a given standard of accommodation. The approach has focused on the choice between rehabilitation of old housing stock and rebuilding (e.g. Needleman 1969, Brookes & Hughes 1975) with emphasis on the purely financial viewpoint of the public agency responsible for housing development. The method may be applied, however, to the broader economic viewpoint of the community.

Needleman devised the seminal model. For the case of a single dwelling, rehabilitation is cheaper than replacement if the cost of rehabilitation, plus the present value of the cost of ultimate rebuilding, plus the present value of the difference in annual running costs and rents for the period of postponed rebuilding, is less than the present cost of immediate demolition and rebuilding, all measured in constant prices. Thus, the cost minimization rule is to rehabilitate if:

$$b > m + b(1 + i)^{-\lambda} + \left(\frac{r + p}{i} \right) \left[1 - (1 + i)^{-\lambda} \right] \qquad (9.3)$$

where b = cost of demolition and rebuilding, m = cost of rehabilitation, i = discount rate, λ = useful life of rehabilitated dwelling (in years), r = difference in annual repair cost between rehabilitated and new dwelling, and p = difference in rent.

Expressing $(r + p)$ as a proportion (α) of (b) and rearranging, the decision rule may be simplified to:

$$(i - \alpha) \left[\frac{1 - (1 + i)^{-\lambda}}{i} \right] > \frac{m}{b} \qquad (9.4)$$

an expression which allows decision makers to determine the maximum acceptable cost of rehabilitation as a proportion of the cost of rebuilding at appropriate levels for the discount rate and future length of life of rehabilitated dwellings. Using sensitivity analysis, Needleman presented results over a range of values for i and λ. It is emphasized that it is important in measuring the various items to keep community economic and local authority financial viewpoints distinct.

Needleman also extended the decision rule to encompass more complicated decision situations. These included a choice between:

(a) complete clearance and rebuilding of a whole area of dwellings on the one hand and partial clearance combined with some modernization on the other; and
(b) rebuilding and rehabilitation where new dwellings are built at a higher or lower density than is afforded by rehabilitation.

He further suggested that the additional annual rent (p) which tenants are willing to pay to live in new rather than renovated accommodation be viewed as providing a crude adjustment for the inherent bias in the decision rule towards rehabilitation, given that no explicit account is taken in the cost effectiveness procedure of the extra benefit tenants may derive from new dwellings built to a higher standard than the minimum requirement. In the application of the model to two projects in the Welsh city of Cardiff, Brookes & Hughes (1975) evaluated this factor by replacing the term ($r + p$) with a term (x), the additional cost of providing in rehabilitated houses the space and condition of the physical fabric available in new houses. For the modified expression indicating that rehabilitation is preferred if:

$$b > m + b(1 + i)^{-\lambda} + x \qquad (9.5)$$

Brookes & Hughes provided the following estimates (in 1972 prices):

$$£7530 > £3816 + £7530(1 + 0.05)^{-30} + £1342 = £6900 \qquad (9.6)$$

Thus, for the two projects under analysis, it appeared that rehabilitation was the better use of available resources.

If it is desired to broaden the analysis still further, then the decision rule may be amended to include consideration of disruption costs involved in rebuilding, as well as perhaps to a lesser extent in rehabilitation. Not that these costs are easily measured. If the approach is expanded yet again to encompass such matters as differential impact on aesthetic amenity and traffic flow, it loses its cost effectiveness character and assumes the identity of full-blown cost–benefit analysis, not to mention the fact that it runs into even more demanding data requirements. It is important, however, that use of the model as outlined by Needleman should not lead to the disregard of these various, less tangible costs and benefits.

9.4 SUMMARY AND CONCLUSIONS

While it is clear that problems of measurement abound in the analysis of urban residential renewal, a CBA framework built on an extension of the basic Rothenberg design, net resident benefits and a cost-effectiveness framework developed by Needleman are available as models for thinking about the social, economic and financial viability of different proposals. This chapter has pointed to the pitfalls involved in the use of these approaches as well as to the possibilities for their useful implementation. It has been shown that some progress has been made, and it has been suggested that more is possible, in extending the estimation of benefits in the Rothenberg model beyond merely site productivity improvements to neighbourhood spillovers and reduced social costs. It has also been shown that as an alternative to measurement of site productivity effects, estimation of resident benefits appears to be feasible using measures of consumer surplus estimated either from formal models or by the contingent valuation approach. Measurement of dislocation costs for displaced residents of former properties may also be approached along the same lines. So far as the cost-effectiveness model is concerned, it has already been officially adopted as the basis for deciding between housing rehabilitation and redevelopment in Britain.

   As a final note, it is to be recognized that matrix display methods outlined in Chapter 6 may also be used for evaluating residential improvement schemes. In the context of urban renewal, these methods have been developed primarily for the evaluation of comprehensive – not merely residential – plans for town redevelopment and expansion. Applications of these methods are reviewed in Chapter 12.

NOTES

1 Note that increased value at the development site refers only to the value of land, not to the value of structures on the land. This is because it is assumed that investment in improvements on the site merely displaces equally productive alternative use of the resources employed so that resulting benefits would have occurred anyway elsewhere in the city. To the extent, of course, that this assumption does not hold, improvement values may be included as benefits (see, for example, Davis & Whinston 1961).

2 Another complication, recognized by Messner 1968, is that since land is valued in terms of the capitalization of expected future returns, private market values are set implicitly on the basis of discounting these returns at a private rather than a social rate of discount.
3 Details of the derivation are provided in DeSalvo 1971.

10

Transportation

In this chapter we assess the use of CBA for evaluating transportation schemes which fall into two categories. The first category comprises infrastructure investments designed to improve inter- and intra-urban and regional transport links in terms of speed, safety and convenience. Examples include construction and upgrading of roads (e.g. urban and inter-urban motorways, city bypasses), railway facilities (e.g. light rapid surface transit, underground lines), ferry systems and bridges. The second category comprises projects of traffic control in cities, examples of which include traffic law enforcement (Shoup 1973) and restraint on automobile use in city centres (Thompson 1967, Gomez-Ibanez & Fauth 1980).

A listing of the principal benefits and costs of infrastructure schemes from the community point of view is given in Table 10.1. For traffic control projects which reduce congestion and delay in the control area – sometimes at the expense of increased travel time, cost and inconvenience to some groups of travellers – many of the same items occur, although they may be costs (benefits) rather than benefits (costs) to certain groups. If a social analysis showing the distribution of effects between different groups in the community is required, transfer payments excluded from the aggregate picture displayed in Table 10.1 are to be taken into account. These include such items as fare revenues, tolls and taxes which would also be highlighted in a purely financial appraisal.

Table 10.1 Benefits and costs of transport infrastructure schemes

Benefits	Costs
user benefits	
vehicle operating cost savings	project resource costs
travel time savings	social dislocation costs
accident cost savings	environmental costs
avoided effort (discomfort and	
inconvenience)	
non-user benefits	
net savings of money, time, risk and	
effort to travellers elsewhere	
economic rent	

Possible effects of a project on income levels in an urban economy through the attraction of new economic activity to the locality are not routinely taken into account in standard evaluation procedures and are excluded from Table 10.1. Such effects are typically regarded as secondary ripples which are remote enough to be overlooked, although if that judgement is not accepted in particular cases, then attempts should be made to include them in transport project appraisals. The general principles of measurement in this regard are not discussed in this chapter, being deferred to Chapter 14 where economic growth effects of transport and other projects are discussed in the context of urban and regional growth and development.

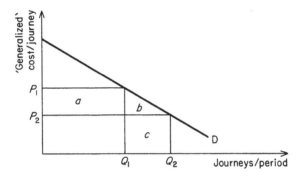

Figure 10.1   Road improvement scheme.

The standard approach towards the measurement of benefits to users of a transport system or facility is in terms of consumers' surplus.[1] Figure 10.1 illustrates the situation for the case of a road improvement scheme. On the vertical axis we plot the so-called 'generalized' cost, or price, of a journey along the route in question, an amount which includes vehicle operating costs (for fuel, maintenance and depreciation) as well as the more intangible costs of time, accident risk and discomfort and inconvenience (sometimes termed effort) as caused, for example, by a poor road surface. On the abscissa we plot the number of journeys made per period. Before the road is improved the 'generalized' cost is P_1 and Q_1 journeys are made. After improvement the cost falls to P_2 because of a better surface and perhaps the elimination of intersections or curves in the road, and now Q_2 journeys are made, additional journeys $(Q_2 - Q_1)$ being 'generated' by the improvement. This information implies a demand function for trips along the road looking something like D. Benefits to travellers who would have travelled the route anyway are represented by the rectangle (a) of cost savings (or additional consumers' surplus). Gross benefits to new travellers are given by the trapezoid ($b + c$) of which the area (c) represents incurred costs. Net benefits to 'generated' traffic are, therefore, given by the area (b). Net user benefits of the scheme are represented by the area of increased consumers' surplus $(a + b)$.[2]

As to non-user benefits, not illustrated in Figure 10.1, the first category listed in Table 10.1 is net consumer surplus gains which may accrue to travellers on other routes or modes to the extent that the 'generated' traffic on the new road is diverted from alternative facilities. There may also be changes in consumers' surplus on complementary facilities if these become more heavily used as a result of the improvement in the system. The principles underlying measurement of consumers' surplus on competing and complementary facilities – termed 'network' effects in this context – have been explained in Chapter 4. Briefly, if there is no change in the price, or 'generalized' cost, of travelling on other facilities, for instance because there is no congestion there, a change in demand for these facilities does not matter. But if the 'generalized' cost does alter, then net changes in consumers' surplus on all facilities in the network have to be taken into account.[3]

Besides 'network' effects, Figure 10.1 also abstracts from another non-user benefit identified in Table 10.1: any change in economic rent, or producer surplus, associated with a project. In an aggregate economic CBA, care must be taken here not to double-count, for user willingness-to-pay may capture the value of certain rent items. For example, any increase in labour income resulting from enhanced job opportunities created by an improved transport link influences, and is reflected in, willingness-to-pay for the improvement. Likewise, any increase in land values occasioned by increased demand for housing in the vicinity of access points to the link indicates merely a transfer of part of consumers' surplus (already estimated) to landowners. On the other hand, other elements of rent are to be included as separate benefits. If a charge is levied for use of a transport facility under analysis (e.g. a toll facility, public transport service or parking facility), profit of operators may be affected.[4]

Empirical implementation of the principles of benefit measurement illustrated in Figure 10.1 presents difficulties in the estimation of 'generated' traffic, the shape of the demand function and measurement of savings in time, accident risk and effort. Reasonable data are generally available in regard to expected vehicle operating cost savings, while estimation of 'generated' traffic is an issue of forecasting which remains a problem in most fields. Here the cost–benefit analyst leaves the problem to the traffic engineer. So far as the demand curve is concerned, it is usually assumed to be linear (not an unreasonable assumption, perhaps, especially if 'generated' traffic is expected to be low). Thus, assuming reliable forecasts, the area of surplus (b) is readily measured. So far as measurement of time savings, accident cost savings and improved comfort and convenience are concerned, we discuss them separately below. We also discuss issues relating to the measurement of social dislocation and environmental externalities, the residual items of Table 10.1 along with project resource costs. Measurement of the latter is straightforward enough in principle not to require detailed treatment. It is understood, of course, that financial costs are to be shadow priced as appropriate for market imperfections, under-utilization of resources, foreign exchange values, indirect taxes and subsidies.

10.1 TRAVEL TIME SAVINGS

Active research into the economic value of time has been going on for at least a quarter of a century and findings have been periodically reviewed (e.g. Watson & Mansfield 1973, Watson 1974, Stopher & Mayburg 1976, Bruzelius 1979). In this section, the plan is not to provide another detailed review of research in the field but to summarize the main issues and findings. The importance of these issues lies in the fact that time savings typically represent the major source of benefits not only in projects designed specifically to reduce congestion, but also in projects providing a general improvement in transport links. Measurement issues relate to the estimation of savings in working time and savings in non-working time. The latter includes commuting time as well as time saved on purely social or recreational pursuits.

10.1.1 Working time

Estimation of the value of working time has been regarded as relatively straightforward. The conventional approach is to measure it in terms of average gross earnings for the type of working time saved, with an additional sum for extra costs associated with the employment of labour (e.g. national insurance contributions by the employer, fringe benefits). This assumes, of course, that time saved can be converted into additional work and output, and that labour is paid the value of its marginal product.[5]

Despite its widespread application, the use of earnings as a proxy for the value of working time involves some obvious difficulties. First, only in a perfect labour market is labour paid the value of its marginal product. If labour creates surplus value as it does in an imperfect labour market, then this surplus in the form of income accruing to other factors of production should also be included in the measure of value. Secondly, this approach ignores the disutility (or utility) of travelling to travellers themselves. Thirdly, it overlooks the possibility of using travelling time productively as, for example, is readily done on train and air journeys.

Another difficulty with the approach is that it is questionable whether the savings of a small unit of time, a minute here or there, can always be put to worthwhile use. On the other hand, it is also possible that the saving of some small amount of time may lead to the acceptance of a longer task which would not otherwise have been undertaken. Moreover, there are many examples of the provision of goods and services designed to save consumers small amounts of time, the market success of which testifies to the fact that significance is attached to small time savings. The result has been that there has been a tendency to allot to small time savings the same value per unit of time as to large time savings. Examples include the Roskill enquiry into the siting of a third London airport (Commission on the Third London Airport 1971) and the guidelines used in the UK for appraising inter-urban road schemes (Department of the Environment 1973). However, recent research confirms that the value of time savings is non-linear: models of the choice of transport mode based on the linear assumption that a minute of time saved is worth the same amount no matter how many minutes are saved appear to overpredict the attraction of such schemes as bus lanes designed to save small amounts of travel time (Heggie 1979). Research is now increasingly pointing to the existence of thresholds of time saving below which savings have lesser, even zero, value.

10.1.2 Non-working time

For the value of non-working time for which there is no obvious market proxy, it is necessary to impute values from choices which people make regarding the use of time (the behaviour observation method) or to use questionnaires or laboratory experiments to elicit a value (the contingent valuation method). It may be noted that if people were free to vary their working hours, the marginal value of non-working time would, in equilibrium, equal the marginal value of working time adjusted downwards for the effect of tax and of the marginal disutility of work. But we should still face the problem of valuing the marginal disutility of work.

Of the two approaches employed, most attention has been given to the behaviour observation method, in particular to the use of time/money cost trade-off models in which implicit time values are estimated from choices between alternative travel modes, routes or speeds. The willingness to pay additional costs in order to save time by a faster alternative provides a measure of the value of time saved. For example, in the following expression:

$$P(A) = a_0 + a_1 dT + a_2 dC \qquad (10.1)$$

where $P(A)$ = probability of choosing route, mode or speed A, dT = difference in time between A and alternative B, and dC = difference in non-time costs between A and B. The ratio a_1/a_2 represents the value of time, being equal to the rate of substitution between time and cost (dC/dT), that is, the amount of cost travellers are willing to incur to save a marginal amount of time.[6] Equation 10.1 may be estimated from cross-section data of trips between pairs of metropolitan zones using multiple regression analysis. It is to be noted that unless data are available for all non-monetary cost differences between A and B, alternatives are required to be identical in all respects other than time and money cost.[7]

On the basis of this trade-off approach there has been a search for a consensus of values so that rules of thumb may be mechanically applied in appraisal studies. Thus from values estimated from studies of commuters' behaviour in the UK in the 1960s, the Department of the Environment (1973) took to using a standard figure of 25% of the average gross hourly wage rate, a figure also employed by the enquiry into the siting of a third London airport (Commission on the Third London Airport 1971).[8]

There are good reasons, however, for scepticism about so simple a procedure because of evidence that the value of time is situation-specific, and the studies from which the value is derived have not taken into account in any comprehensive way all the various situations which are relevant. For one thing, commuters appear to dislike waiting and walking more than time spent in a vehicle (Quarmby 1967), so that there is a case for distinguishing between types of travelling time saved. Again, the value of time appears to vary with trip length, a given amount of time being more important on a short than on a long trip (Hensher 1976). There are also different types of leisure time saved, and it is not surprising that social-recreational time has been found to be valued differently to commuting time saved (Watson 1974). Most behaviour observation studies have analysed the value of commuting time only. Finally, the value of time has been found to vary positively with income (e.g. Beesley 1965, Quarmby 1967, Hensher 1976), so that the value of time savings depends on who makes the trips in question. At the very least, then, if an overall average value is to be used for non-working time savings, it should ideally be based on findings stratified according to type of travelling time saved, trip length, trip purpose and income class. Such a requirement involves much more extensive investigation than has been carried out to date, although the UK Department of the Environment does now distinguish between the time of different types of travellers (car drivers, bus passengers, car passengers, bus drivers, etc.) reflecting different income levels. The Department also distinguishes between in-vehicle leisure time and walking and waiting time.

These difficulties aside, other complications arise in connection with behaviour observation studies. First, these studies assume perfect information on the part of travellers regarding time and money cost alternatives. Secondly, observed estimates of time reflect differences between alternatives in features other than money cost and time (e.g.

comfort and convenience, scenic attractiveness). Thirdly, the problem of small time savings arises again.

Before leaving the behaviour observation approach, it is worth mentioning that estimates of the value of time have been attempted from observations on property values (e.g. Mohring 1961, Wabe 1971). Using multiple regression analysis, variations in property prices are explained in terms of the various characteristics of property, including distance (time) from different facilities, notably the central business district. The value of time saved in commuting, shopping, and so on is then measured in terms of the additional price paid for accessibility, after standardizing for other influences on property price. This method, the standard hedonic price method described in Chapter 5, suffers from the shortcomings discussed in that chapter.

As an alternative approach to the behaviour observation method for valuing non-working time savings, the contingent valuation approach has been much less frequently used. However, in light of a growing awareness of the several difficulties with the more popular approach and of widely varying values resulting from it, both of which factors cast doubt on the reliability of the single averages traditionally used in empirical studies, greater attention is beginning to be given to the alternative method. While the possibility of bias from hypothetical responses and other sources remains, the method may not be without merit, as we have suggested in Chapter 5. Two recent studies illustrate the use of the approach. Hensher (1972) employs the questionnaire survey method, asking people how much extra money they would be willing to pay before changing their chosen mode of commuting for a slower mode. Hauer & Greenough (1982) employ a laboratory test consisting of offering subjects cash in return for the loss of specified amounts of time in order to gauge the compensation required for loss of time. Neither study should be viewed as more than exploratory, but each suggests a promising approach to the valuation of time.

10.2 ACCIDENT COST SAVINGS

Improvements in transport systems are invariably expected to yield reductions in accident rates or in the severity of accident effects. In these cases the analyst is required to estimate expected savings in terms of avoided property damage, death and injury. Measurement of the value of property damages is relatively straightforward. Measurement of the value of averted death or injury has proceeded in terms of avoided cost, defensive expenditures and the statistical value of reduced risk. We examine the worth of each of these approaches in turn.

10.2.1 Avoided costs

The avoided cost approach involves the measurement of output and income not forgone as a result of system improvement, together with the avoidance of property (mainly vehicle) losses, medical and police expenses. In some applications, these avoided losses are augmented by an apparently arbitrary allowance for avoided pain, grief and suffering on the part of both victims and others.[9] The cost of lost output and income avoided is measured in terms of the present value of expected gross earnings (sometimes, surprisingly, less the individual's own consumption). Thus, ignoring the consumption adjustment, the value (V) of earnings not forgone as a result of reducing the incidence of death or injury is defined as:

$$V = \sum_{t=1}^{n} \sum_{i=1}^{m} P_a P_b E_{it} (1 + r)^{-t} \tag{10.2}$$

where P_a = probability of individual i surviving through year t, P_b = probability of individual i being employed in year t, E_{it} = gross individual earnings in year t, r = social discount rate, $t = 1 \ldots n$ years of earnings saved, and $i = 1 \ldots m$ individuals affected.

Although this approach has been widely used in the past, there are numerous criticisms of it. First, it ignores the probabilistic element in individual calculation. Schemes to reduce death or injury are aimed at reducing risk rather than avoiding the death or injury of specific individuals. Thus benefits are better measured in terms of the value of reduced risk (see Section 10.2.3) than in terms of numbers of deaths and injuries avoided. Secondly, the approach ignores the intrinsic value, over and above the value of earnings, of living; and of living a life free of injury, except to the extent that an allowance is made for pain, grief and suffering. Thirdly, the approach implies that the lives of non-earners such as housewives, volunteer workers and retired people are valueless, unless some attempt is made, using either market proxies or contingent valuation, to put a value on household and volunteer production. This is not usual. Fourthly, the approach ignores the avoided loss of non-working time or leisure (Thomas 1978), and fifthly, if own-consumption is deducted from the earnings stream, the provocative implication is that society is rendered better off by the death of retired people whose consumption exceeds earnings. The consumption benefit of his/her earnings to the individual concerned is also ignored. Finally, as always in the case of the use of earnings as a benefit proxy, the assumption is that labour is paid the value of its marginal product.

10.2.2 Defensive expenditures

Two possibilities have been explored in connection with imputing values for the cost of death or injury from expenditures incurred to avoid these misfortunes. The first possibility is to examine past decisions by society on such expenditures as highway safety, search and rescue, and medical research for implicit values of life and safety. For example, suppose a safety barrier on the central reservation of a motorway is estimated to save 20 lives during its economic life at a present value installation and maintenance cost of $4 million, then a life is implicitly valued at $200,000. The problems with this method are that it fails to yield values within even a tolerably narrow range, results are complicated by purely political reasons for certain expenditures (e.g. highway expenditures designed to win elections) and it begs the question in regard to estimating the returns to projects for the improvement of highway safety, search and rescue facilities, and so on.

The second possibility is to impute values from insurance coverage purchased. Thus the value of life might be estimated on the basis of the amount of life or flight insurance the 'typical' individual buys.[10] One problem here is that individual perceptions of risk, which as indicated earlier are what should really underpin valuation, are not likely to be perfect. Another problem is that the amount of life and flight insurance purchased reflects a concern for the welfare of others rather than the utility that an individual places on his/her own life. Thus, as Mishan (1982a) has observed in one of his characteristically florid passages, 'a bachelor with no dependants may have no reason to take out flight insurance, notwithstanding which he could be as reluctant as the next man to depart this fugacious life at short notice' (p. 325). In the case of injury insurance, the approach may have somewhat more merit.

10.2.3 Statistical value of reduced risk

There is a growing awareness that a conceptually more satisfactory way of addressing the issue of accident loss savings is to measure them in terms of the willingness to pay for a reduction in the probability of death or injury. This is aside from avoided medical, police and property damage costs as well as avoided losses of net output (gross output less own consumption) which measure victims' contributions to society.

The basic idea can be illustrated for a single individual as in Figure 10.2, where one indifference curve from a whole map of curves is depicted, showing combinations of wealth (W) and the probability of death/injury (P) which yield equal utility. The task is to evaluate dW/dP from a given transportation proposal for a population of individuals in order to estimate the so-called statistical value of life (freedom from death) or freedom from injury, that is, the value of a defined change in risk (SV). Thus:

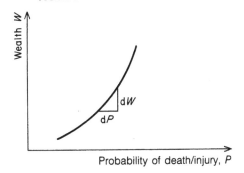

Figure 10.2   Risk trade-off curve.

$$SV = \sum_{i=1}^{n} dW_i/ndP \qquad (10.3)$$

where dW_i = willingness-to-pay of the ith individual for a given reduction in the probability of death or injury, n = population of individuals affected, and P = probability of death or injury. The term ndP is the expected number of deaths or injuries avoided so that if, for example, mean $dW_i = \bar{W}$ = \$5 for dP = 1/100,000, then SV = \$500,000.

Given that engineering forecasts are likely to be available for dP, the empirical problem is to measure $\bar{W}$. The hedonic price and contingent valuation methods are available. Using the hedonic price method, attempts have been made to gauge the willingness-to-pay for reduced risk from labour and consumer good markets where risk differences are observable. Thus, Thaler & Rosen (1975), using the multiple regression method, explained wage rate differences between occupations in terms of differentials in skill levels, market conditions and occupational risk. The positive coefficient showing the change in wage rate with respect to a unit change in risk provides an estimate of the mean reduction in compensation accepted for a reduction in risk. Needleman (1980) estimated the statistical value of life, not from a cross section of occupations, but from bonus pay for accepting additional danger in construction work. As examples of the imputation of an implicit price from consumption activity, Blomquist (1979) and Ghosh et al. (1975) derived estimates of the time cost or price people were willing to pay to secure a reduced probability of death from, respectively, seat belt use and slower travelling speeds.

Difficulties with the hedonic price approach are that it requires adequate control for determinants of wage rates or consumption practices other than the characteristics on which the analysis focuses. In addition to this requirement, it is also highly demanding in terms of data regarding probabilities and, in the case of the Blomquist and Ghosh et al. studies, time values. It further assumes full information on the part of workers or consumers of the relevant probabilities. Moreover, results will vary according to circumstances (cause of death or type of injury), which fact argues against the simple use of an average value of life or freedom from injury in all circumstances. Results also abstract from the value of reduced risk for other parties who are spared the cost of pain, grief and suffering. Finally, results from labour market studies are not likely to be representaive of willingness-to-pay for risk on the part of the population as a whole inasmuch as implicit values are derived from the preferences of a group of untypical risk-takers who accept the riskier jobs.

The contingent valuation approach, though it has so far received much less attention than the hedonic price method, avoids most of the foregoing difficulties. On the other hand, it can be costly to implement, especially if it is deemed necessary to ask people what they are willing to pay to secure reduced risk both for themselves and, separately, for others. It is also subject to the possibility of bias resulting from hypothetical responses and the other sources discussed in Chapter 5. A major recent feasibility test directed at estimating the value of life, however, appeared to show that there is no reason to be concerned about the willingness of people to respond to questions, that responses display reasonable consistency, that subjective estimates of risk probabilities are reasonably reliable in the transport context and that there does not appear to be any systematic non-representation of true preferences (Hammerton et al. 1982).

In light of these findings from small-scale laboratory tests, a major survey has been mounted in the UK to derive estimates of the statistical value of life for transport risks (Jones-Lee et al. 1985). The questionnaire contained questions falling into three categories: valuation questions intended to provide estimates of the value of fatal and non-fatal accidents; perception/consistency questions designed to test the quality of individual perceptions of transport risks, ability to handle probability concepts and rationality of choices; factual questions concerning such matters as annual mileage driven, age and income that would influence individual valuations. A follow-up survey was conducted to test for the temporal stability of responses, and a mean value of statistical life for transport risks was estimated of $£1.5 \times 10^6$, or about $£2 \times 10^6$ if willingness-to-pay for others' safety as well as one's own was taken into account. On the basis of the perception/consistency and stability tests employed, results were considered to be quite robust, and, by comparison with other recent estimates based on labour market data (Marin 1983), not unreasonable.

10.3 COMFORT AND CONVENIENCE

There have been few attempts to measure separately the comfort and convenience of travellers as these may be improved by a new transport link or as they vary between alternative transport proposals. Differential measures of the value of time in different activities (driving, walking, waiting, being driven), however, as used in standard appraisal procedures, partly reflect differentials in the amount of effort (the degree of discomfort and inconvenience) involved in each activity.

Goodwin (1976) argues that attempts should be made to measure effort separately since it is logically distinct from time in the 'generalized' cost of travel and is capable of independent manipulation over time. He suggests three broad approaches. The first is to use an ordinal points system regarding different travel activities so that some consideration can be given to differences in comfort and convenience even though monetary measurement may not be possible. Thus Hensher *et al.* (1974) used the interview technique to establish point scores; alternatively, proposals may be ranked in terms of the number of travel speed changes involved in journeys. The second, doubtless more ambitious approach is to employ some physiological measure of effort expended: for instance, kilo calories expended per unit of time, heart rate or galvanic skin response to emotional strain. The third approach is to attempt to place a monetary value on saved effort.

An interesting, if less than satisfactory example of the monetary evaluation method occurred in the appraisal of the new Victoria underground railway line for London (Foster & Beesley 1963). The enhanced comfort and convenience for both travellers on the new line and travellers remaining on the existing underground system was measured in terms of the value of the increased probability of obtaining a seat at peak hours. It may be noted that any benefit to travellers on other transport modes who would have also gained from the diversion of traffic to the new line was ignored, as were dimensions of reduced effort (e.g. number of transfers required per journey) other than the probability of not having to stand en route.

At any rate, assuming that the valuation of the change in the probability of getting a seat is proportional to this probability change as well as to distance travelled (both of which assumptions are at least questionable), an implicit value was derived from the trade-off of comfort and convenience against time on slow and fast trains. For the marginal passenger who is just persuaded to take a slow train, this value (C) is defined as follows:

$$C = \frac{V(t_s - t_f)}{(D_f t_f - D_s t_s)} \tag{10.4}$$

where V = monetary valuation of commuting time, t_s = travelling time on slow train, t_f = travelling time on fast train, and D = disutility of travelling = $100 - x$, where x = probability of obtaining a seat. Travel times were valued in terms of results from previous studies and probabilities of obtaining a seat from the intuition and experience of railway staff. Results were then adjusted by arbitrary amounts to take account of the fact that non-marginal travellers on slow trains as well as travellers on fast trains would also be willing to pay some unknown higher price for increased comfort and convenience.

It is not surprising that the method has not been replicated in precise detail, given the limited scope of the measure of reduced effort, some of the heroic assumptions employed and the casual empirical procedures adopted. Yet it suggests a line of approach which, with considerable refinement, may yield more useful findings.

10.4 SOCIAL DISLOCATION AND ENVIRONMENTAL EFFECTS

While these effects are rarely evaluated in transport projects, they may be of substantial significance in certain cases, notably in urban motorway and urban bypass schemes.

10.4.1 Social dislocation costs

The costs of social dislocation have already been discussed in connection with housing renewal schemes (Ch. 9) so that it is not necessary to examine the issues in detail here. It is sufficient to mention that the following costs may be involved: the monetary costs of removal of households displaced by a scheme; their lost surplus which may be estimated using the contingent valuation approach (Commission on the Third London Airport 1971); losses to those remaining as neighbourhood residents in the form of severed personal relationships, visual intrusion and increased inconvenience regarding pedestrian and vehicle access to homes and local facilities. The latter effects may be directly measurable using contingent valuation and, in certain cases, indirectly measurable using time valuation methods. Delays to traffic during project construction may also be relevant as a form of social dislocation.

10.4.2 Environmental effects

In addition to effects in terms of social dislocation, road construction schemes also have a physical impact on the environment in terms of the possible creation, both during construction and later, of noise, vibration, fumes, visual disamenity, the loss of historical buildings or other artefacts, the loss of wildlife habitat and the loss of open space for recreational purposes. Other proposals such as restraining traffic volume in cities have positive effects in these terms. Of procedures available for measuring intangible costs and benefits as discussed in Chapter 5, four are commonly considered for measuring pollution effects and other environmental impacts: the damage cost, defensive expenditure, hedonic price and contingent valuation methods.[11] But, in truth, little progress has yet been made in building such measures routinely into transport appraisals.

One way of measuring the value of environmental impact is in terms of the cost of environmental damage created by, say, a road proposal or relieved by, say, a scheme to restrict auto use in a city centre. Thus it may be possible to predict from past experience the effect of the scheme on the wear and tear, and hence repair and maintenance costs, of such materials as zinc, steel, rubber, concrete and stone. Likewise for an urban bypass proposal, it may be possible to estimate the likely effect on the yields of commercial crop and livestock farming. So far as health damage is concerned, it is necessary to estimate damage functions along the lines pioneered by Lave and Seskin (1970, 1977).[12] The incidence of different diseases is explained on the basis of a range of determinants, including environmental pollution variables. Thus, a function of the following form might be estimated using multiple regression analysis on a sample of either time-series or cross-section data:

$$M_i = M(A, Y, D, \ldots Q) \tag{10.5}$$

where M_i = morbidity or mortality rate for disease i, A = age structure of the population, Y = measure of income distribution, D = density of population, and Q = environmental quality. From the partial derivative $\delta M_i/\delta Q$ we obtain a measure of the impact of environmental quality on illness or mortality rates from which, using procedures discussed in Section 10.2.3 for valuing life and freedom from illness, it is in principle possible to estimate the health cost of a scheme causing estimated degrees of noise or air pollution.

Difficulties, of course, abound in connection with the estimation of damage functions. For one thing, results may be acutely sensitive to model specification, that is, to the functional form chosen (linear or non-linear) and to the range of independent variables included. Errors of observation also present a problem. Disease-specific death rates, for example, are susceptible to error due to misclassification, partly as a result of relatively few autopsies being conducted. Moreover, different measures of air and noise pollution are available. All in all, there is a need for much more extensive research into damage functions than has been undertaken to date, before this approach is to be used with much confidence.

An alternative approach to the valuation of environmental damage is to use proxies in the form of expenditures incurred for the purpose of mitigating the effects of pollution or damage to historic or natural assets. One well known example of the use of this approach was the use of insurance values by the Roskill enquiry (Commission on the Third London Airport 1971) to give what was surely an underestimate of the social cost attached to the loss of historic churches to make way for a new London airport. Another example, discussed in Chapter 5, aimed to estimate the economic value of peace and quiet from expenditures on sound-proofing in houses near London airport (Starkie & Johnson 1975).

The hedonic price method has received widespread application in the context of measuring environmental damage (e.g. Anderson & Crocker 1971, Walters 1975). From a cross section of observations in a metropolitan area, the method is used to explain differences in property prices in terms of various characteristics of properties such as size, accessibility, natural amenities and exposure to pollution of various kinds (Q). Thus, as outlined in Chapter 5, the model may be written as follows:

$$P = P(Q_1 \ldots Q_n, H_1 \ldots H_m) \tag{10.6}$$

where P = house price, Q = vector of neighbourhood environmental characteristics, and H = vector of other housing characteristics. The marginal implicit price of each type of pollution (i) is defined in terms of the partial derivative $\delta P/\delta Q_i$, the increase in expenditure necessary to obtain a unit reduction in noise level or air pollution.

Despite extensive experimentation, the method must be judged to offer only tentative guidance regarding the cost of pollution in particular circumstances, given the several difficulties associated with its use. These, again, have been discussed in Chapter 5; briefly, they relate to the assumptions underlying the approach, problems of model specification and problems of data reliability and availability. A recent review of hedonic price studies concerned with highway noise and property values, however, concludes that main highway noise reduces the price of an 'average' home by 8–10%, and that this range may be usable in cost–benefit studies (Nelson 1982).

The final method available for gauging the environmental impact of transport proposals is contingent valuation. As in other applications of the method, it appears to deserve more serious investigation than it has so far received, although the problems of bias remain. In the environmental context, for example, it may be questioned whether individuals are competent to make meaningful valuations when it is difficult to discern clearly the hazards of, say, toxic air pollutants. It may also be difficult to elucidate from individuals the full value of environmental costs when some of those costs fall on public authorities (through health services and insurance schemes) and on other individuals in the form of grief and suffering to friends and relatives. Even so, bearing these difficulties in mind, a usable average or range of value for specific types of nuisance such as traffic noise from urban motorways may be built up as more evidence accumulates from field studies.

In summary, then, there are in principle methods available for measuring the social and environmental effects of transport schemes, yet they are fraught with difficulties in application. Extensive further research is required to improve the results obtainable from these methods. In the meantime, there is a sufficient choice of methods to allow very approximate valuations, or ranges of value, to be pieced together for particular proposals. It is certainly not appropriate to ignore the effects in question, as has so often been the case.

10.5 CASE STUDY ILLUSTRATION

An example of an attempt to take account of most of the costs and benefits of transportation projects previously discussed is provided by Pearce & Nash (1973) in a revision of an original evaluation of a proposed urban motorway for the city of Southampton. Taking account of 'network effects', the original study estimated savings in vehicle operating costs, travellers' time and accident costs at approximately £6 million (1971 prices) for the first year after project completion (1981) on the basis of standard formulae provided by government departments. However, these annual 'generalized' cost savings were applied to forecasted traffic *with* the motorway, some proportion of which would not have travelled in the absence of the motorway (i.e. would be 'generated' traffic) and, assuming a linear demand curve, would derive per unit benefits valued at 50% of the cost savings to other units of traffic. Moreover, according to Pearce & Nash (1973, p. 138), it was not clear that accident cost savings, a minor component of total cost savings, would be significantly different from zero, given that less traffic would travel the system in the absence of the motorway. Assuming that (a) 25% of forecasted traffic would be 'generated', and (b) accident cost savings were zero, Pearce & Nash put the 1981 benefit figure nearer to £3.6 million than £6 million for users of the network of roads affected by the motorway. It is to be noted that neither the original study nor the Pearce & Nash revision included any value for reduced discomfort and inconvenience for travellers.

On the cost side, the original study was said to have underestimated the true opportunity cost of capital and maintenance outlays. It also failed to make any allowance for intangible social costs, including dislocation and environmental effects. Capital and maintenance costs were undervalued in various ways. First, the costs of complementary interchange investments were omitted from the cost of the motorway. Secondly, publicly owned land used for the motorway scheme, along with property on that land, was undervalued in terms of alternative market value. Thirdly, recorded construction costs were assumed to be understated, given that land schemes typically turn out to be more costly than estimated. Fourthly, certain infrastructure facilities would need to have been built to accommodate residents displaced by the motorway. Finally, some of the residents displaced by the scheme decided to move some time before compulsory purchase orders were issued, causing under-utilization of the housing stock since their properties remained vacant or were demolished earlier than necessary. Notional rent for the period of 'waste' should, therefore, have been debited to the scheme. Making these adjustments, Pearce & Nash estimated project resource costs (discounted at 10% to 1972 values) at around £20 million (in 1971 prices), approximately twice the value of recorded historical capital and maintenance costs.

Table 10.2 Motorway social costs
(discounted present values, 1971 prices).

	£m
extra cost of rehousing	0.27
lost householders' surplus	1.32
disruption during construction	0.92
traffic delay during construction	0.76
noise nuisance to householders	0.26
unquantified costs	u
	$3.53 + u$

Source: Pearce & Nash 1973, p. 138.

In terms of intangible costs, Pearce & Nash provided conservative estimates, as shown in Table 10.2. The first item, extra cost of rehousing, refers to the movement cost incurred by householders, estimated at £1000 per owner–occupier. Householders' surplus represents the difference between owners' subjective valuations of homes and their market price, estimated at 52% of market price following a Roskill Commission finding (Commission on the Third London Airport 1971) based on contingent valuation research. Disruption during construction refers to the noise imposed on residents, an externality valued by assuming that the 3500 houses affected would suffer a 30% reduction in annual nominal rental value for at least two years. Delays to traffic during construction were valued on the basis of an average delay of five minutes per trip for 60 000 trips for one year as estimated from data for 1969 traffic flows. Motorway noise nuisance for residents who remained in the vicinity of the new road was estimated on the basis of likely insulation costs (using OECD data) necessary to achieve a noise climate of 60 dBA or less. Finally, unquantified costs included such items as severed personal relationships, visual intrusion and loss of open space and other recreational facilities.

Table 10.3 Motorway evaluation: revised results
(discounted present values, 1971 prices).

	At 3% p.a. growth in benefits (£m)	At 5% p.a. growth in benefits (£m)
costs	$23.13 + u$	$23.13 + u$
benefits before completion	2.52	2.52
benefits after completion	21.27	23.28
net present value	$0.66 - u$	$2.67 - u$

u = Unquantified costs.
Source: Pearce & Nash 1973, p. 141.

Revised results are shown in Table 10.3 under alternative assumptions concerning the annual growth in benefits. Bearing in mind that intangible costs were estimated on a conservative basis and recognizing that certain costs remained unquantified, the motorway scheme appears marginal at best. It would be for decision makers to decide whether, in their judgement, the unquantified costs would be significant enough to outweigh the measured net benefits. As Pearce & Nash point out, consideration should also be given to alternative ways (e.g. public transport improvements) of achieving transportation objectives, and to the distributional effects of the scheme. Many of the costs would be borne by lower income groups, while beneficiaries would be car drivers, a large proportion of whom would be from higher income groups.

10.6 MATRIX DISPLAY METHODS

In view of the numerous problems involved in estimating monetary values for all the possible effects of transportation schemes, a valuable approach may be considered to be matrix display, which provides an indication of the extent of intangible impacts through non-monetary measures where available. A simple result expressed in net present value, benefit–cost ratio or internal rate of return terms is not derived; instead the several effects of proposals – in this context time, accident rate, human effort, social dislocation and environmental impacts – are displayed in matrix form and decision makers are left to exercise judgement in the weighing of monetary versus non-monetary considerations and of the relative importance of the different non-monetary effects. An additional advantage of this approach is that it also shows the distribution of impacts as between different sectors. The two best-known examples of this type of appraisal of transportation proposals are the Planning Balance Sheet (PBS) and Goals Achievement Matrix (GAM) methods. As explained in Chapter 6, the GAM was developed explicitly for application to transportation proposals. Examples of the use of the PBS in this field include evaluations of alternative relief road proposals for the town centres of Edgware (Lichfield & Chapman 1968) and York (Lichfield & Proudlove 1976). For detailed discussion of the use of matrix display methods, see Chapter 12.

10.7 SUMMARY AND CONCLUSIONS

Considerable effort has gone into the refinement of transportation project analysis over the years, with notable progress being made in procedures for estimating the key items of vehicle operating cost savings, travel time savings and accident cost savings. However, the need for still further effort remains, especially in respect of time and accident costs. In the search for average values of time for use in everyday appraisals, perhaps the central issue in connection with time research, it is recognized that a single value is inappropriate; that time has to be valued according to at least a few different types of time saved (e.g. working time, non-working time, in-vehicle and waiting or walking time, time on long and short trips). It is also beginning to be recognized that the contingent valuation method could offer some promise as an alternative to the more traditional time/cost choice method for valuing time. In the matter of accident cost saving, greater attention is being given to the statistical value of reduced risk as opposed to lost income and defensive expenditures as measures of avoided death or injury.

So far as measurement of comfort and convenience, social dislocation and environmental effects are concerned, less progress has been made. Again the contingent valuation method appears to offer some hitherto relatively unexploited promise.

NOTES

1 There have been attempts to use impact on national income rather than surplus as the measure of benefit (e.g. Bos & Koyck 1961, Tinbergen 1957), but one of the main limitations of that aproach is that it fails to take into account intangible effects – of major significance in transport projects – unless an output proxy can be found for their valuation. In the absence of externalities, shadow pricing and changes in factor inputs, it can be shown that changes in welfare correspond to the mean of changes in national income measured on the basis of Paasche and Laspeyres indices (Pearce & Nash 1981, pp. 99–102).
2 Use of the ordinary Marshallian demand curve as the basis from which to measure consumers' surplus assumes, of course, the absence of an income effect (see Ch. 4).
3 There is a path dependency problem here if the marginal cost of travelling on the improved road is less than perfectly elastic. This means that the measure of surplus depends on the sequence of price changes on the different roads unless income elasticities of demand for each facility are equal. The absence

of this requirement is not usually regarded as likely to bias results very seriously in practice (see Ch. 4).

4   An item given consideration in some treatments is net change in indirect tax revenue for the government. If tax revenues (T) are affected by a switch in expenditure between transport and non-transport sectors, then any such effect (dT) may be regarded as akin to a producer surplus and, as such, should be included in the analysis. Measurement is not particularly complex if it is assumed that total expenditure in all sectors as well as average indirect tax rates by sector remain constant so that:

$$dT = dE_t(i_t - i_n)$$

where E_t = expenditure in the transport sector and $i_{t,n}$ = average indirect tax rates in transport and 'other' sectors. It is a judgement as to whether this is a matter important enough to be considered in a particular appraisal.

5   This approach is not confined to explicit labour market activity. The value of housewives' production (and hence time) has also been estimated by applying market wage rates to their job as disaggregated into its various work categories (e.g. Weisbrod 1961).

6   For constant $P(A)$, $a_1 dT = a_2 dC$. Thus $a_1/a_2 = dC/dT$.

7   For other methods of estimating the time/cost trade-off, see Beesley 1965, Watson & Mansfield 1973.

8   This figure was also used in Australia and New Zealand. In the USA, during the 1970s, a figure of 42% of the average gross hourly wage rate was widely used (see Hensher & Truong 1985).

9   A variation on this approach is to refer to court awards regarding death and injury because these awards are designed to compensate for loss of income as well as some grief and suffering.

10 If, for example, the risk of being killed on an air trip is 0.00025% and the individual pays one dollar in flight insurance, he/she may be viewed as being willing to pay one dollar to reduce the risk to zero, implying that the individual values his/her life at $1.00 \times 100/0.00025 = \$400,000$.

11 We discuss procedures for valuing historical sites, wildlife and open space by the so-called travel-cost method in Chapter 11.

12 For a recent example, see Small 1977.

11

Recreation

In both urban and regional planning contexts, information on economic and social as well as purely financial benefits and costs is important in decisions regarding the provision of recreational facilities: the need for them, their size and number, their optimal use levels, their locations and the timing of their provision. Such facilities include multi-purpose urban leisure centres, swimming pools, theatres, libraries, golf courses, sports pitches, urban and wilderness parks, historic homes, wildlife reserves and so on.

Focusing on the aggregate community point of view, benefits and costs may be conveniently classified as in Table 11.1. If a social analysis is required, effects may be disaggregated according to socio-economic, ethnic or geographical groups. If a purely financial analysis is needed by public officials, benefits and costs respectively appropriated or incurred by government may be separated out.

Table 11.1 Benefits and costs of recreation facilities

Benefits	Costs
user benefits	project resource costs
non-user benefits	site congestion costs
economic rent	ecological damage costs
option value	social inconvenience costs
existence value	(e.g. noise, road congestion)
reduced external social costs	
(e.g. crime costs, health costs)	

It may be necessary to explain the meaning and relevance of some of the effects listed in Table 11.1. Option value refers to the amount which current non-users would be willing to pay to preserve a recreation area or facility in case they may wish to become users some time in the future. Existence value refers to the willingness-to-pay on the part of non-users for the satisfaction of knowing that the recreational opportunity exists, even if they themselves never expect to enjoy it. Thus people may be willing to protect a remote wilderness park from mining or logging development in the interests of protecting the natural ecosystem for its own sake or, at any rate, for the benefit of others. Congestion costs created by crowding at the site are relevant in the cases of the *ex post* evaluation of an existing facility and the analysis of optimal use level.

In what follows in this chapter we discuss the measurement of effects listed in Table 11.1. It will be obvious that not every project necessarily generates all of the effects listed in the table. Moreover, as in the case of urban renewal and transport projects, recreation projects may stimulate local economic growth and development, notably in this case by attracting visitor spending to an area. If analysis is conducted from the point of view of the local economy, such dynamic effects should be included as aggregate benefits. Consideration of these matters is again left to Chapter 14.

11.1 USER BENEFITS

The benefits to users of recreational facilities are to be measured in terms of consumers' surplus following the principles outlined in Chapter 4.[1] Two methods have been used for this purpose, the travel-cost and contingent valuation methods, which we now discuss in turn.

11.1.1 *Travel-cost method: basic model*

The travel-cost method uses travel costs incurred in getting to the recreation site as the basis for estimating a demand curve for the facility. Although there have been different approaches to the use of travel costs, the so-called Clawson–Knetsch method has dominated the field (Clawson & Knetsch 1966).[2]

The Clawson–Knetsch method involves a two-stage procedure. At the first stage the demand curve for 'the whole recreational experience', or the visit generation function, is estimated from data on visits/thousand population per period from different zones around the site. This is the relationship between cost/visit (including travel cost and entrance price) and visits/thousand per period. At the second stage the demand curve for the 'recreation site *per se*', the relationship between entrance prices and visits per period, is estimated by postulating hypothetical increases in entrance charges, given the assumption that a unit change in entrance fee elicits the same response in terms of visits as does a unit change in cost/visit. Thus if an extra dollar of total visit cost (from zone B as compared with zone A) means that 100 fewer visits/thousand are made from B than from A, then a hypothetical increase of one dollar in the entrance charge would mean 100 fewer visits/thousand.

The method may be implemented by constructing '*per se*' demand curves for each zone, computing consumers' surplus for each zone and aggregating over zones, or by constructing an aggregate '*per se*' demand curve by estimating total visits from all zones at each entrance price and computing aggregate consumers' surplus as the area under this aggregate curve above the actual entrance charge. Abstracting from non-user benefits (see below), the value of this surplus/period may then be compared with operating costs to gauge the economic value/period of a recreation resource. If information is available on the alternative use value for the site (e.g. residential construction value), the relative worth of the recreational service may also be judged. A clear illustration that estimates the value of a trout fishing lake is provided by Smith (1971). Seeley (1973) provides illustrations of the recreational value of such facilities as an urban park and a golf course relative to site value for housing development.

A simple numerical and graphical example may clarify the basic approach. Suppose for simplicity that all zones have equal population size and that entrance price for the facility is zero. Figure 11.1a shows the demand curve for 'the whole recreational experience', Figure 11.1b the 'demand curves *per se*' for each zone, and Figure 11.1c the aggregate 'demand curve *per se*'. Aggregate consumers' surplus for the recreation site in question is computed as $3750/period from the sum of areas under the curves in Figure 11.1b, or from the area under the curve in

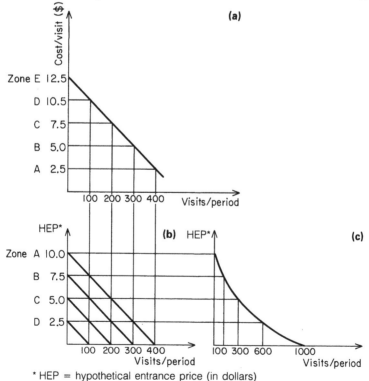

* HEP = hypothetical entrance price (in dollars)

Figure 11.1 Clawson–Knetsch model.

Figure 11.1c. Also notice that since the curve in Figure 11.1a is common to all zones, consumers' surplus may alternatively be computed directly from the demand curve for 'the whole recreational experience' as the sum of areas above each cost/visit and below the demand curve (Burt & Brewer 1971, Mansfield 1971). This fact helps to simplify computations in practice.

11.1.2 Travel-cost method: assumptions

In order to judge the usefulness of the basic model, it is worthwhile to examine the assumptions on which it rests. We suggest that, although empirical difficulties abound, the assumptions of the method are not such as to deprive it of practical value.

The assumptions of equal population zones and zero entrance price were made for expositional convenience and may be easily abandoned without altering the logic of the model. In order to take account of zonal variations in population size, the visits variable may be defined in unit population terms; or else population size may be included as an independent variable in the visit function.[3] The assumption of linearity in the demand curves was also made for convenience. In practice, the chosen form of the visit generation function is often quadratic, log-linear (a power function) or semi-log (an exponential function), reflecting the fact that the model is a special case of the general gravity model of social interaction. By contrast with these assumptions, however, the notion that the response to a change in entrance price is identical to the response to a change in total visit cost is crucial to the validity of the model. If this is not the case, it becomes necessary to estimate the elasticities of demand with respect to these different prices. In fact, practitioners tend to accept the assumption as not unreasonable in most circumstances.

Other assumptions involved in the Clawson–Knetsch model have remained implicit so far. First, use of an ordinary Marshallian demand curve for estimating consumers' surplus implies a zero income effect (see Ch. 4). This assumption is generally justified on the ground that recreational expenditures typically represent a small proportion of individual budgets so that the income effect would be negligible. Secondly, the graphical example developed in Section 11.1.1 assumed homogeneous zones of origin in terms of all characteristics other than distance from the site. We have already dealt with population size. But zones will also differ in respect of other factors likely to influence zonal demand for recreation: income, the accessibility and attractiveness of competing recreational opportunities, and various sociodemographic characteristics (age structure, ethnic composition) affecting preferences for the type of recreation in question. In practice, these other influences may be taken into account in the fitting of the visit generation function, using multiple regression analysis to standardize for their effects. Thus in one well known formulation (Cesario & Knetsch 1976), the function takes the following form:

$$V_{ij} = \theta X_i Y_j \exp(\beta C_{ij}) K_i^{\alpha} \tag{11.1}$$

where V_{ij} = visits from zone i to site j, X_i = vector of population characteristics of zone i, Y_j = vector of characteristics of site j, C_{ij} = travel cost from i to j, K_i = index of accessibility/attractiveness of k sites for population at i

$$= \sum_{k=1}^{M} Y_k \exp(\beta C_{ik}) \tag{11.2}$$

and $\theta.\beta.\alpha$ = estimated parameters. Standardization for variables other than travel cost and population size is important if any of these variables are correlated at all with travel cost or population. If they are, bias in coefficients resulting from omitted variables occurs in the absence of standardization.

A third implicit assumption of the model is that travel to the recreation site yields, of itself, no benefits. This may be a reasonable assumption in many cases, particularly so far as urban recreational facilities are concerned. But in other cases visitors may derive positive utility from getting to the recreation site if the travelling is seen partly as an end in itself (e.g. recreational motoring). In these cases, the model will overestimate the value of a facility unless travel costs are reduced in accordance with travel benefits derived. One approach is to use questionnaires to distinguish between 'pure visitors' who secure site benefits only and 'meanderers' who secure both site and travel benefits (Cheshire & Stabler 1976), and then to remove from the meanderers' travel costs some proportion of time costs involved. Sensitivity analysis may be used here. Another approach available for evaluating travel benefits is the contingent valuation method.

A further implicit assumption is that time costs are included with vehicle operating costs in the measure of travel cost. Omission of time costs, as occurred in early studies, biases the demand curve downwards from its true position and thus leads to underestimation of site benefits (Cesario & Knetsch 1970, Cesario 1976). The easiest procedure to account for the cost of time is to add in estimates of the value of leisure time made in previous studies.[4] Thus Cesario (1976) and Cesario & Knetsch (1976), for example, use a range of between one-quarter and one-half of average wage rates on the basis of accumulated research findings to the mid-1970s. Unfortunately, we have seen in Chapter 10 that some reservations attach to the use of standard average values of this sort.[5]

The travel-cost model also assumes that only one recreational site is visited per trip. If this is not the case and several sites are visited during a trip, it is necessary to prorate travel costs across sites in accordance with the marginal cost of visiting each site. Otherwise, the standard application will overestimate site value. Unfortunately, this requirement involves additional questionnaire detail at the data collection stage to identify what Cheshire & Stabler (1976) termed 'transit visitors' together with their trip patterns. The model also assumes that a single tariff applies for entry to the recreational site. In many cases, in fact, optional tariffs apply, users being given the choice between a price per visit or a season ticket involving a lower average price per visit. Gibson & Anderson (1975) have pointed out that standard application of the model underestimates benefits in the presence of the season ticket option if a weighted average of per visit charges is used. These authors, therefore, stress the need for separate estimates for season ticket holders and day visitors. Moreover, the method assumes that people do not move home to be nearer a recreational site (Anderson 1974b). If they do, the method understates the value of the site, and increased property values resulting from such movements should, strictly speaking, be added to travel cost benefits. It is not, however, likely that this effect introduces significant distortion since relocation probably seldom occurs.

Finally, the basic model assumes that efficiency alone is important, giving no explicit recognition of the relevance of distributional concerns. Thus development of a ski slope for the benefit of suburban car owners prepared to travel considerable distances is likely to yield a higher net surplus than are facilities for the urban poor. But that is not to say that the ski slope should necessarily receive priority. It is always possible to introduce differential weights on zonal surpluses to reflect distributional judgements, bearing in mind the difficulties of establishing such weights (Ch. 6). At least it should be recognized that decision makers must ultimately exercise discretion – in favour of the disadvantaged if they wish – using the outcome of an aggregate economic analysis as informational input rather than as conclusive evidence regarding desirable resource allocation.

11.1.3 Travel-cost method: system capacity adjustments

So far we have looked at the measurement of user benefits for a facility viewed in isolation from the system in which it exists. For an evaluation of a unique recreational service this is an adequate perspective. In many empirical applications, however, interest lies in the value of additions to, or reductions in, the capacity of a recreational system. In cases of additions, for example, it is necessary to take account of the partial or total displacement of existing facilities offering the same type of service (Vickerman 1974, 1975). This means that the differential rather than total willingness-to-pay has to be gauged of users who are diverted from existing facilities, as well as the willingness-to-pay of users who are 'generated' by development of a new facility or extension of an existing one. In what follows we discuss the details of this issue with reference to provision of a new facility, but point out where the principles also apply to the case of extending an existing facility.

One way of addressing this issue in practice if the proposed new facility is identical to an existing one (excepting that it is less costly to reach from some zones) is to use visit data for the existing facility to estimate consumers' surplus for each facility assuming non-existence of the other. The difference in consumers' surplus represents the value of the new facility. If this exceeds its costs measured in comparable annual or present value terms, then it should be built. This is the analogue of the case of a new road development explained in Chapter 10. A simple example is given by Williams & Anderson (1975, pp. 43–5) of the case of replacing a single central library with new suburban libraries which lower the price of library services for each zone in a metropolitan area from, say, P_0 to P_1 (Fig. 11.2). For each zone, the gain from having a suburban library system (abstracting from the likelihood that the range of books immediately available will be lower with suburban libraries) is represented by the areas (*a*) and (*b*), where (*a*) measures the gain in surplus accruing to diverted borrowers (those who would have used the old library anyway) and (*b*) measures the gain to 'generated' or new borrowers. Assuming non-user benefits to be negligible, a suburban library system would be developed if the present value of these user benefits aggregated over zones exceeds the present value of capital and operating costs associated with the new system.

An illustrative variation on this theme concerns the value of establishing a US national air museum in Los Angeles in addition to, rather than in place of, the existing museum in Dayton, Ohio (Swaney &

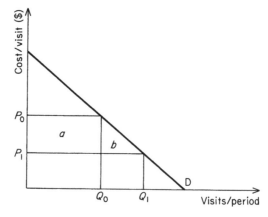

Figure 11.2 Library services.

Ward 1985). Using data from an on-site survey of visitors to the existing museum during the summer of 1982 (non-summer visits were arbitrarily ignored), a linear demand function for 'the whole recreational experience' was estimated on the basis of (a) trips/thousand population from each of 60 zones and (b) standardization for influences on visits other than travel cost. Based on distance from the largest population centre in each zone, travel cost included vehicle cost valued at \$0.27/mile as well as time cost valued, using information from another study, at 60% of the average national gross income from work. Consumers' surplus was then estimated for each zone and aggregated across zones for both the existing museum and for the two-museum system (Table 11.2). In the two-museum case, people from zones nearer to Los Angeles than to Dayton derived additional consumer surplus from two sources: the reduced trip cost for those who would have travelled to Dayton anyway and the welfare gain for 'generated' visitors induced to visit the museum given the lower trip cost. Results denote a very substantial consumer gain from the new facility. It is to be noted as well that by evaluating alternative sites rather than merely a single new site, the procedure outlined could be extended for use in the context of an optimal location decision regarding the new facility.

Table 11.2 Estimated annual consumers' surplus (\$000 1982) from existing museum and two-museum system (selected origin zones only)

Origin zone	Consumers' surplus existing museum	Consumers' surplus two-museum system
Arizona	–	1 907.03
California	0.26	35 184.10
Colorado	0.04	22.48
Indiana	293.29	293.29
Michigan	61.86	61.86
Oregon	–	1 663.60
.	.	.
.	.	.
.	.	.
All zones (rounded)	21 000.00	65 000.00

Source: Swaney & Ward, 1985, p. 178.

In the library and museum examples just discussed, the new facilities were assumed to provide identical services to those provided by the original facilities. It is more usual, however, that services of proposed facilities are not identical to those of existing ones so that a common demand curve is not applicable. An approach for dealing with this problem of partial homogeneity between facilities was developed by Mansfield (1971) and used by Anderson (1975). A common demand curve was employed but only a proportion of diverted benefits was counted in the measure of value of the new facility. The determination of the appropriate proportion, of course, remains arbitrary unless questionnaire surveys are used to establish numbers of those who would divert (in an *ex ante* study) or did divert (in an *ex post* study).

An alternative procedure is to work with the demand function for the whole recreational system, say, golf courses in a metropolitan area or national parks in a region. The impact of any increase in supply, whether new facilities or extensions of existing ones, may then be evaluated in terms of surpluses for users diverted from alternative facilities and for users newly generated for the system. In both examples of increased supply, access price falls as a result of reduced congestion in the system, and in the case of new facilities as a result of increased accessibility. Doling & Gibson (1979) tested this method in respect of urban swimming pools.

Still another alternative is to begin with a visit generation and distribution model from which may be estimated demand functions for each facility in the system. The Cesario–Knetsch (1976) formulation given in Equation 11.1 is one such model. It can be re-written as follows to show the generation and distribution components more clearly:

$$V_{ij} = \left[\theta X_i K_i^{(\alpha+1)} \right] \left[\frac{Y_j \exp(\beta C_{ij})}{K_i} \right] \tag{11.3}$$

The symbols are as defined earlier. The term in the first square bracket is the generation component showing the number of visits from zone i to facility j as a function of origin characteristics (X_i) and accessibility of all facilities (defined in terms of both attractiveness and cost of access). The term in the second square bracket is the distribution component showing that the number of visits from i which terminate at j is determined by the ratio of the accessibility of j to the accessibility of all facilities in the system. Introduction of a new facility j effectively increases accessibility to that facility for people in zone i from zero to $Y_j \exp(\beta C_{ij})$, increasing system accessibility (K_i) by the same amount. Given calibrated values for θ, α and β, the effects throughout the system may be predicted.

The question now is how to value the change in consumers' surplus attaching to the establishment of the new facility. The aggregate demand curve for the new site is estimated from Equation 11.3 using hypothetical values for C_{ij} and summing over i. Following the principles outlined in Chapter 4, the change in surplus resulting from the new facility depends on whether or not prices at substitute facilities fall as demand for their services recedes. If not (perhaps because zero price is charged anyway) any apparent 'loss' of surplus at these facilities is irrelevant, such 'loss' being merely the result of users switching to the new facility where they perceive themselves to be better off. The value of the new facility is, therefore, the area under the aggregate demand curve for j and above the price line (which could coincide with the base axis). If, on the other hand, the user price of other facilities does fall (perhaps because they were previously overcrowded and the cost of congestion falls), then it is necessary to add to surplus at the new facility any gain in surplus resulting from reduced prices at existing facilities, and any further gains at new and existing facilities resulting from continued feedback effects if these are important. For this purpose it is necessary to estimate new demand curves for new and existing facilities from the generation–distribution model.[6] It may be noticed that these same considerations apply to the case of capacity extension at one facility (which reduces congestion there and increases its attractiveness) and that the visit generation–distribution model can again be used as a basis for estimating net gains.

11.1.4 Travel-cost method: other considerations

Before leaving the travel-cost method it is worth mentioning that it has shortcomings other than those associated with its assumptions. First, use of a log specification of the trip generation function presents a minor difficulty when it comes to benefit estimation because the function is asymptotic to the vertical axis, implying that benefits are infinitely large. And only arbitrary procedures are available to circumvent this. Mansfield (1971), for example, adds a constant to the dependent variable (visits/ population) to give zero visits at a finite travel cost. Common (1973) suggests the judgemental imposition of a cost maximum as the upper limit of integration. Secondly, it has been pointed out that user response to a fall in supply price as a result of increased system capacity may be larger than trip generation models calibrated to existing use levels predict (Vickerman 1974). This is because improved attractiveness of services offered may lead to considerable revision of leisure habits. In this event, the standard procedure will underestimate the net benefits of system expansion.

A third drawback is that the method is not useful for evaluating facilities when travel costs are negligible and do not appear as significant determinants of use. This may be particularly the case in respect of recreational services in small urban areas, although tests of the model in respect of, *inter alia*, beach parks on the island of Oahu in Hawaii (Moncur 1975) and urban forest parks in Chicago (Darragh *et al.* 1983) have confirmed the usefulness of the approach in such a context. Fourthly, it has been shown that the traditional use of visit data highly aggregated by zones introduces a negative bias into estimates of travel cost elasticity (Flegg 1976). This suggests that excessive aggregation should be avoided, even though it has the effect of improving the overall fit of the model, because there is a lower variation of visits among groups of zones than among individual zones. Lastly, the method can be quite costly to implement as a result of the need to develop trip generation and often, ideally, trip distribution models.

11.1.5 Contingent valuation method

As in other areas of cost–benefit application, the contingent valuation method has received relatively little attention in the recreation field, although its advantages are increasingly being recognized. It is not necessary at this point to reiterate all the pros and cons of the method as discussed elsewhere in the text. Examples of its use in gauging recreationists' benefits include attempts to measure the willingness to pay for and/or the compensation required to forgo access to waterfowl hunting (Hammack & Brown 1974) and fishing opportunities (Sinclair 1976), to recreational beaches (McConnell 1977) and to wilderness activities in general (Chicchetti & Smith 1973). In the latter two examples, the method was at the same time used to estimate the negative impact of site congestion on willingness-to-pay. McConnell, for example, estimated consumers' surplus/visitor day at a sample of Rhode Island beaches as a function of income, visits/season, air temperature and degree of congestion. The method was to ask respondents for merely a 'yes' or 'no' answer regarding beach attendance, and to increase price hypothetically by 50¢ until a 'no' answer was received. It was estimated that, *ceteris paribus*, an extra 100 people on a beach reduces consumers' surplus/visit by about 25%.

11.2 NON-USER BENEFITS

As Table 11.1 indicates, there are community benefits of recreation projects to be taken into account in addition to user benefits. These include producer surplus, option and existence values, and, conceivably, reduced external social costs in the form of crime and ill health. It is also possible that recreation developments may generate additional local income by attracting visitors from outside the jurisdiction whose viewpoint frames the analysis. In this section we discuss the estimation of these additional benefits with the exception, as pointed out earlier, of income generation effects.

If the marginal cost curve of a recreational facility is upward-sloping to the right rather than horizontal, or a positive entrance charge is made above marginal resource cost (whether or not marginal cost is constant), then an element of producer surplus or economic rent is to be included in total benefits (Smith 1975). To the extent that a price in excess of marginal cost is charged, the supplier simply appropriates part of what would otherwise have been consumers' surplus.

In addition, land and property values close to recreation sites may increase, providing another source of economic rent to be counted in addition to benefits measured from travel costs. We have already made mention in Section 11.1.2 of the need to add to travel cost benefits any enhanced property values resulting from relocation by users in order to be nearer a recreational facility. Here, however, we are concerned with willingness-to-pay for a general amenity on the part of non-users, benefits which may be significant in connection with proximity to, say, urban open space and parks. In practice, both user and non-user benefits reflected in enhanced property values are likely to be estimated together. Two methods of estimation are available for this purpose, the difficulties involved in each having been discussed already. The Rothenberg approach, outlined in Chapter 9, is one method whereby property prices before and after development of a facility are observed, taking due account of such other influences on price as the effects of inflation and changing supply conditions. The other method is the hedonic price approach whereby land and property prices are explained in terms of property characteristics, including in this context proximity to recreational facilities. An example of the use of the hedonic price method to gauge the value of urban water parks is given by Darling (1973).

So far as option and existence values, defined at the start of the chapter, are concerned, the best method for estimating them would appear to be contingent valuation. (For illustrations of the use of contingent valuation for estimating both types of value in respect of wildlife preservation and public parks, see Brookshire *et al.* 1983 and Majid *et al.* 1983 respectively.) A tentative alternative procedure for estimating option value is mentioned in Section 12.2.2 below (Schwind 1977).

In connection with reduced social costs, it is expected that improved recreational facilities will enhance standards of health and fitness, and may also reduce the incidence of crime, but studies of the extent to which these effects involve resource cost savings (over and above benefits to individual recreationists themselves) are still awaited. One way to proceed is to estimate damage functions as discussed in Chapter 10, and to attach to results of reduced health and crime damage estimates of resource cost savings involved. In principle, contingent valuation or *ex post* statistical evidence from elsewhere may also be considered. It should be noted that considerable research effort may be required for what could be relatively insignificant effects. The balance of benefits and costs connected with this effort is, therefore, to be considered carefully.

11.3 COSTS OF RECREATIONAL FACILITIES

Referring again to Table 11.1, we list relevant costs as the resource costs of a recreational project together with external costs comprising site congestion, ecological damage and social inconvenience. Estimation of the resource costs of service provision is relatively straightforward, bearing in mind the need to shadow price resources used (in particular, in this context, land) and to take account of risk and uncertainty as appropriate.

In appraising the *ex post* value of a facility, or in determining optimal use level, it is necessary to take account of congestion costs at the facility. In either an *ex post* or an *ex ante* study of outdoor recreation projects, and again in determining optimal use level, it may be necessary to estimate costs of ecological damage to the countryside. It may also be appropriate to take account of such other external costs as road congestion or neighbourhood noise and general disturbance caused by recreationists. There has not been substantive progress yet in putting a monetary value on these various social costs but the contingent valuation method offers some promise. It has already been used to value the costs of site congestion in monetary terms (e.g. Fisher & Krutilla 1972, McConnell 1977). An alternative way of coping with both ecological damage and site congestion costs is to maximize net benefits (excluding these costs) subject to physical constraints reflecting tolerable levels of these externalities. This approach was adopted in a study designed to establish optimal use levels for a wilderness park (Walter & Schofield 1977), constraint values being left to decision makers to formulate. Finally, the contingent valuation, hedonic price or defensive expenditures approaches, as discussed in Chapter 5, may all be used in principle to estimate the cost of inconvenience to non-users of a facility.

11.4 SUMMARY AND CONCLUSIONS

The cost–benefit method can be applied to a range of issues in the area of recreational decision making. The most widely used method for estimating user benefits – the Clawson–Knetsch travel cost method – involves numerous assumptions and requires considerable research effort. But the assumptions are not generally fatal to its value in that some may be inconsequential in many circumstances and others may be removed in practice, albeit at the expense of still further research effort. The variety of applied studies available bears testimony to the worthwhileness of that effort. So far as non-user benefits are concerned, considerable research effort is required to estimate such items as option and existence values, and reduced social costs; and such benefits could be relatively insignificant in many situations. As to costs, there remain assorted, if not insuperable, difficulties of estimation. But, as with benefit estimation, progress is being made all the time.

NOTES

1 Primary and secondary classes of users may be distinguished. While primary users represent the market for which a facility exists, secondary users derive an incidental benefit from its existence. For example, primary users of a golf course are golfers; secondary users are passers-by or walkers who secure a beneficial non-golfing experience from the existence of the course.

2 For an example of a generally discarded alternative method, see Pearse 1968, with follow-up comments by Brown & Nawas 1972, Pearse 1972 and Gibbs 1974.

3 Rather than use of visits/population as the dependent variable, it has been suggested that population size be included as an independent variable in order to avoid the implicit assumption that the response of visits with respect to population is unity (e.g. Common 1973, Flegg 1976).

4 Cesario & Knetsch 1976 also experiment with a multiplicative relation between money cost and time cost in estimating the generalized cost of travel.

5 As an alternative, Common 1973 recommended an iterative procedure for estimating the marginal valuation of time spent travelling from visit data.

6 For the case when many prices change simultaneously, it is necessary that income elasticities of demand for all facilities be equal (homothetic utility functions) in order to obtain a unique welfare outcome (see Ch. 4). Any actual divergence from this requirement is unlikely to be very important in practice.

12

Comprehensive land-use planning

Land-use planning goes by various names in addition to the one chosen here: town and country planning; town planning; physical planning; or spatial planning. Whatever the name, its objective is to optimize the use of land as it is taken up for different purposes: shopping and other commercial development, industrial development, residential development, agricultural, transportation or recreational use. Separate projects within these categories of land use may, of course, be subject to cost–benefit appraisal as we have seen in the three previous chapters where the principles of housing, transportation and recreational project analysis were discussed. In this chapter we focus on the evaluation of broader plans for urban redevelopment and expansion, and for spatial development at the regional level, where a variety of interrelated projects may be involved.

Use of CBA in the spatial planning field has been dominated by matrix display techniques in recognition of the multitude of economic, social, political and physical considerations associated with the location, form and effect of urban and regional development. However, less comprehensive methods have also been used, ranging from the strict monetary measurement of aggregate welfare change used in conventional CBA to mere financial appraisal. In this chaper we review in turn the matrix display and more partial approaches to plan evaluation.

12.1 MATRIX DISPLAY METHODS

Most prominent among matrix display methods in the area of land use development have been the Planning Balance Sheet (PBS) and Goals Achievement Matrix (GAM) mentioned in Chapter 6. Other frameworks, however, have also been developed; and after illustrating the application of PBS and GAM, we outline a recent alternative, the so-called Evaluation Matrix for Decision Makers (EMDM; Poulton 1982).

12.1.1 Planning Balance Sheet

Lichfield's PBS attempts to indicate the extent of all community impacts of proposals whether measurable in monetary units or not. In the absence of monetary measures, physical units of measurement are employed where available, or else costs and benefits are included qualitatively. At the same time, the method displays the distribution of impacts between different sectors of the community, classified according to whether they are producers/operators who play a part in creating and running the services involved in the plan, or consumers of the services produced. Results are laid out in a set of detailed accounts (the balance sheet) with decision makers left to weigh up the relative importance of the various incommensurable effects and the interests of the different sectors in the process of attempting to maximize aggregate welfare.

Several case studies using the PBS method for appraising alternative city development plans are available (e.g. Alexander 1974, Lichfield 1966a, 1969, Lichfield & Chapman 1968, 1970). The method has also been extended to the evaluation of larger-scale regional and subregional plans (for an assessment of selected examples, see Lichfield *et al*. 1979). We illustrate the approach by reference to the case of the proposed expansion of the town of Ipswich (Lichfield & Chapman 1970). Though dated, this case remains a typical example of the genre. Five alternative schemes involving encroachment on to surrounding agricultural land were evaluated. Table 12.1 lists the sectors of the community affected. Costs and benefits are defined as if development would take place all at one time so that no discounting is involved. At the summary stage of the analysis, a qualitative assessment is made as to whether there would be any material difference in time streams of benefits and costs between proposals for which discounting would alter conclusions.

Table 12.1 Ipswich expansion: sectors affected

Producer/operators	Consumers
development agency	public in the expanded town
current landowners	in town and district centres
displaced by plan	in principal residential areas
not displaced	in principal industrial areas
local authorities and ratepayers	in principal open space and recreational areas
	on principal communication systems
	vehicle users
	public transport
	pedestrians
	current occupiers
	displaced
	not displaced

Table 12.2 Modified planning balance sheet: Ipswich expansion schemes

Sector	Possible unit of measurement		Schemes					Objectives	Order of preference: greatest benefit or least net cost				
	Cost	Benefit	I	II	III	IV	V		I	II	III	IV	V
Producers/ operators													
Development agency		M	£19 000+ $M_1 - M_7$ Year 1	£31 000+ $M_1 - M_7$ Year 1	£29 000+ $M_1 - M_7$ Year 1	£25 000+ $M_1 - M_7$ Year 1	£25 000+ $M_1 - M_7$ Year 5	Low net financial cost	0	−2	−1	−1	−1
								Early development of the land	0	0	−1	−2	−3
								Further urban growth	0	−2	−2	−2	−3
Reduction									0	−4	−4	−5	−7
Current landowners													
Displaced													
In urban and village areas	m	M	n	n+	n+	n++	n+	Nil net cost	0	0	0	0	0
Agricultural landowners and farmers	m	M	n++++	n+	n+++	n++	n	Minimum annual loss agricultural output	0	+3	+1	+2	+4
Reduction									0	+3	+1	+2	+4
Not displaced													
In urban and village areas	i		n++	n+	n+	n+	n	Choice of service facilities	0	−1	−1	−1	−2
		n	n	n	n	n	n	Environmental amenity	0	0	0	0	0
Reduction									0	−1	−1	−1	−2

Sector							Measure					
Farmers	m	n+++	n+	n+	n++	n	Minimum annual loss agricultural output	0	+2	+2	+1	+3
Local authorities and ratepayers	m	£203+	£301+++	£299+++	£285++	£276/294 ++++	Municipal cost	0	-4	-2	-2	-1	-3
Producers/operators overall reduction								0	-4	-4	-4	-5
Consumers												
The public in the expanded town							Choice of facilities	0	0	0	0	0
In the town and district centres:							Location: (a) proximity to other centres	0	-2	-2	-3	-1
							(b) accessibility to west side	0	-1	-1	-2	-3
Commercial occupiers and users of public/private buildings	m	i	n	n	n	n	Environmental amenity	0	-1	-1	-2	-3
Reduction								0	-4	-4	-7	-7
In principal residential areas:	m	i					Minimum severance by roads	0	-1	-1	-2	-1
Remaining or new residents		n++	n+	n	n		Minimum aircraft noise	0	-1	-1	-1	-1
		n+	n	n+	n+		Environmental amenity	0	-2	-2	-1	-3
Reduction								0	-4	-4	-4	-5

Note: Money measures in thousands.
Source: Lichfield & Chapman 1970, pp. 184–5.

Table 12.2 shows a segment of the balance sheet, containing information on the impacts of each scheme on all producers/operators and on selected groups of the public as consumers of services in the expanded town. Although examples of each type of measure are not shown in the table, units of a capital nature are identified by upper case letters in the columns headed 'Cost' and 'Benefit' as they might be measured in units of money (M), time (T) and other physical description (P) or remain intangible (I). Annual flows of benefit or cost are denominated in lower case letters (m, t, p, i).

In the columns headed 'Schemes', scale of impact is recorded in relation to each of the presumed group objectives shown in the column headed 'Objectives'. The letter 'n' refers to numbers of individuals involved. Unquantified capital costs (in seven categories) are assigned a subscripted upper case M (a subsidiary analysis showed ordinal rankings in terms of unquantified capital costs). The plus signs indicate orders of magnitude. In the columns headed 'Order of preference', the projects are ranked in comparison with scheme I according to each group objective (plus signs now denote preferred ranking relative to scheme I). In these last columns, rankings are also summed to show in 'reduction' rows the ranking of proposals from the viewpoint of each group.

It is not necessary to describe each entry in detail. However, a few points of clarification are in order. Benefits to displaced landowners are in the form of compensation while costs represent foregone net revenues. Since compensation is based on foregone net revenues, the proposals are seen to be neutral in terms of net cost. In the case of farmers not displaced by the schemes, loss of agricultural output results from interference in farming by the increased population accommodated by the schemes. We may perhaps question whether minimization of agricultural output loss for the nation should be included as an objective of farmers (both displaced and not displaced).

Results of the full analysis, of which Table 12.2 is a part, are displayed in Table 12.3, which shows the relative merits of proposals from the standpoint of each group. From the bottom line of the table, it appears that scheme I is the most desirable overall. However, from the particular point of view of farmers (including consideration of lost output for the nation) and some vehicle users other schemes would be preferred. It is for decision makers to weigh the advantages and disadvantages of the different projects before coming to a final choice.

12.1.2 Goals Achievement Matrix

The GAM was developed by Hill (1968) for specific application to transportation proposals, although it has been adapted for evaluation of city and regional development plans (see Lichfield *et al.* 1979). Costs and benefits are arranged according to community goals (e.g. regional

Table 12.3 Planning balance sheet summary:
Ipswich expansion schemes

Sector	Schemes				
	I	II	III	IV	V
Producers/operators					
Development agency	0	−4	−4	−5	−7
Current landowners					
Displaced					
In urban and village areas	0	0	0	0	0
Agricultural landowners and farmers	0	+3	+1	+2	+4
Not displaced					
In urban and village areas	0	−1	−1	−1	−2
Farmers	0	+2	+2	+1	+3
Local authorities and ratepayers	0	−4	−2	−1	−3
Producers/operators overall reduction	0	−4	−4	−4	−5
Consumers					
The public in the expanded town					
In town and district centres:					
Commercial occupiers	0	−4	−4	−7	−7
In residential areas: remaining or					
new occupiers	0	−4	−4	−4	−5
In principal industrial areas:					
industrialists and workers	0	0	0	0	0
Users of regional open space and					
countryside	0	−1	−1	−1	−1
On principal communications					
system					
Vehicle users: Internal traffic	0	−2	−6	−4	−3
External traffic	0	+1	0	−1	+2
Through traffic	0	+1	−1	0	0
Public transport: external	0	0	0	0	0
Pedestrians	0	0	0	0	0
Current occupiers					
Displaced					
In urban and village areas	0	−1	−1	−1	−2
Agricultural occupiers			not certain		
Not displaced					
In urban and village areas	0	−1	−1	−1	−2
Agricultural occupiers	0	+2	+2	+1	+3
Consumers overall reduction	0	−9	−16	−18	−15
Producers/operators and consumers overall reduction	0	−13	−20	−22	−20

Source: Lichfield & Chapman 1970, p. 188.

economic growth, noise reduction) as well as groups affected. The main difference from the PBS is that goals and group interests are explicitly weighted to reflect their relative importance in the community welfare function, a characteristic which exposes the method to criticisms of prior weighting as outlined in Chapter 6.[1] An example is provided in Table 12.4 for the case of a single plan.

Table 12.4 Abbreviated goals achievement matrix

Incidence	Goal a, relative weight 2			Goal b, relative weight 3		
	Relative weight	Costs	Benefits	Relative weight	Costs	Benefits
Group a	1	A	D	5	E	–
b	3	H	}	4	–	R
c	1	L	J	3	–	S
d	2	–		2	T }	–
e	1	–	K	1		U
		Σ	Σ			

Source: Hill 1968, p. 23.

The letters A, B . . . are costs and benefits that may be defined in monetary or non-monetary units, or in terms of qualitative states. Recognizing the role of sensitivity analysis, Hill sees these as probably representing a range of values. A dash means that no cost or benefit relating to an objective accrues to the group in question, while certain impacts indicated by braces, cannot be disaggregated as between groups. Where possible, because they are expressed in commensurable units, costs and benefits are summed with respect to goals. There is no explicit distinction made between capital and current items. But use of the equivalent annual value method (Ch. 3) to convert the capital value to equivalent annual values enables measurement of all effects to be conducted in comparable annual terms. Again, it is for decision makers to weigh the various advantages and disadvantages of alternative plans.

12.1.3   Evaluation Matrix for Decision Makers

A feature of both the PBS and GAM is that they present a welter of information, often in somewhat confusing array. The EMDM was proposed by Poulton (1982) as a simpler approach, designed to focus attention on elements judged to be most significant for the final decision. Table 12.5 presents a truncated application involving analysis of alternative uses of a waterside site in Vancouver, Canada. Although five alternatives were considered in the case study, only two are illustrated here: use as a park and use for a large-scale hotel, office, retail and apartment development. Moreover, external impacts on non-users of the site, although included in the original analysis, are not shown in the table.

The EMDM consists of four matrices showing the impact of alternative schemes on community values, specific land-use objectives, the city treasury and community groups affected. Cell entries in italics indicate that an element is judged to be of high importance, while roman entries are of low importance.

12.1.4 Matrix display models: assessment

The advantages of matrix display models are that they attempt to incorporate formally into the analysis the multitude of intangible effects associated with development plans and to show the distribution of effects between different groups in the community. On the other hand, they usually depart from the welfare economics foundation which underlies conventional CBA. This is because numerous benefits and costs are not typically measured in terms of welfare surpluses through the notions of willingness-to-pay and compensation required.[2] Pure financial or physical measures of impact are used instead. Moreover, no attention is given to the issue of shadow pricing market items on the cost side. The result of these features, together with the emphasis on distributional effects, is that the aggregate economic efficiency dimension of plans tends to be lost, doubtless the consequence of the development of models out of a tradition of physical planning rather than economic analysis.

At the same time, models depend heavily on the exercise of judgement, not only by decision makers, which is unavoidable, but by analysts as well. In principle, judgements regarding the choice of relevant community and group goals could be framed in consultation with decision makers and/or through a process of community consultation. In practice, however, analysts usually accept a high degree of responsibility in setting goals. Judgements regarding differential weights have also in practice been exercised by analysts, although in principle consultation should be possible and sensitivity analysis is always available for use. Finally, professional judgements are used in the choice of appropriate measures for gauging the extent of non-monetary benefits.

12.2 ECONOMIC COST–BENEFIT ANALYSIS

The methodology of conventional economic CBA has found widespread application in the planning of developments in particular sectors of the urban and regional economy (see earlier and subsequent chapters). It has also been applied in limited, preliminary fashion to issues of land use planning addressed in this chapter; namely, the alteration of land-use patterns for purposes of town centre redevelopment and town expansion. The advantage of conventional CBA is that it attempts to highlight the

Table 12.5 Evaluation matrix for decision makers: Harbour Park

Item	Commentary	Park only	Large-scale development
Values			
fairness	General values exerting influence on final decision [1]Concern with unusual confiscation of property rights	–	site zoning changed
equity	[1]Transfer of benefits from few people to many or vice versa	many → few (park acquisition)	few → many (revenues to city)
Objectives	Synthesized by City Planning Dept, reflect all major concerns		
Public use	[4]*% area in public use*	*100%*	*20%*
yacht marina	[4]number of berths provided	350	?
view at ground level	[4]% frontage blocked at ground level	0	80%
transport improvement	[3]Flow capacity and quality increase in excess of generated demand	+	–
no hotels, offices, apartments	Anti-development position expressed as constraint	yes	no

Fiscal objectives			
Break-even on commercial operations	Revenue to the city; ⁵Revenue from operation of marina and developments covers costs (min. position)	$30 000 p.a. surplus on marina	large surplus
All costs covered	⁵All operating and site preparation costs covered; substantial contribution to site acquisition costs	$252 000 p.a. deficit on operations and site preparation costs	large surplus
Affected parties: site users	Includes all groups on behalf of which strong pleas have been made		
Park users	²Park is small and disturbed by traffic, excellent substitutes are available	slight benefit	
boat owners	³Rent and services increased, net effect beneficial to some	slight benefit	
apartment owners	²Buyers obtain net benefits that exceed those from other options		benefit – possibly large
users of commercial developments	Hotel and retail customers, office and shop renters		benefit – possibly large from use of view location

Note: Code numbers attached to items indicate degree of quantification used: 1, concern named; 2, direction of change indicated; 3, ordinal scale used; 4, physical quantities used; 5, money values used.
Source: Poulton 1982, pp. 94–5.

aggregate economic dimension of schemes, although it typically provides less comprehensive measurement of community impacts than attempted by matrix display methods. Indeed, the empirical problems associated with monetary evaluation of the varied impacts of multi-project development schemes have served to limit its usefulness to date. In this section we illustrate application of conventional CBA separately to schemes for urban redevelopment and town expansion.

12.2.1   Urban redevelopment

Examples of the use of conventional CBA to analyse schemes for urban redevelopment include assessments of alternative proposals for the South Bank area in Central London (Flowerdew & Stannard 1967, Flowerdew 1971) and of the redevelopment of the old Covent Garden market area of London (LeFevre & Pickering 1972, Kirk & Sloyan 1978). It indicates the difficulty of implementing conventional CBA in this field that none of these studies succeeds in measuring many relevant items in monetary terms.[3]

Drawing on the lessons of earlier chapters, a listing of the principal benefits and costs of redevelopment from the point of view of the urban community affected is given in Table 12.6. As in previous chapters, we leave for later consideration (in Ch. 14) the possibility that schemes may generate economic development which would not have otherwise occurred. For a social analysis, benefits and costs may be disaggregated according to income, geographical or other groups affected.

Table 12.6   Benefits and costs of urban redevelopment

Benefits	Costs
increased site productivity	project resource costs
neighbourhood spillovers	social dislocation costs
reduced public protection service costs (crime, health, fire, traffic accidents)	loss of urban character
site user benefits for non-occupiers*	
reduced congestion costs	
reduced accident costs	
reduced crime costs	
recreational benefits	

*Not already included in preceding benefits.

Increased land values as suggested by Lean (1967) may be used in principle to measure the benefit of improved productivity at the site of redevelopment and beneficial spillovers on to neighbouring properties, as in the case of residential urban renewal. Increased land values reflect specific benefits for which occupiers at the site and neighbouring sites reveal a willingness-to-pay. Local government may capture part of the increased land value in increased tax revenues, although this item is merely a transfer from the point of view of the community as a whole. Specific benefits accruing to site occupiers and occupiers of neighbouring properties include such items as profit opportunities for commercial operators, net environmental pollution and visual amenity improvements, reduced congestion and accident costs, travel cost savings connected with reduced commuting and other travel distances to the extent that redevelopment schemes create additional living accommodation within the city, reduced risks of fire, disease and crime, and enhanced recreational opportunities.[4]

There may be doubt, however, that sufficiently sensitive forecasts of land values can be made to reflect differences in the quality of alternative plans, especially where large tracts of land are dedicated to non-revenue producing development. The problems of estimating altered land values standardized for increases resulting from influences other than schemes under analysis, and for offsetting reductions in value elsewhere in the city, have been emphasized in Chapter 9 regarding residential renewal projects. For larger, multi-project urban redevelopment schemes such problems are compounded. For sceptical views of the prospects of using land value estimates for measuring benefits, see Lichfield (1968, 1970) and Evans (1969). In the absence of reliable land value estimates, it is necessary to estimate separately the various benefits which land value changes reflect, using contingent valuation, hedonic prices and/or surrogate market approaches where market values do not exist. While possibilities for the use of these approaches exist as indicated in earlier chapters, the difficulties associated with them are not to be underrated.

In addition to land productivity effects, redevelopment may involve benefits for persons other than occupiers of site premises and neighbourhood properties as indicated in Table 12.6. Savings in real resource costs may be secured in the provision of public protection services, while site users who are not occupiers may enjoy benefits in terms of reduced congestion, accident and crime costs, and improved recreational opportunities. With the exception of crime costs, the possibilities and problems of placing a monetary value on these items have been discussed in previous chapters. So far as crime costs are concerned, it is necessary to consider property loss (to the extent that assets leave the community through the process of crime), property damage, injury costs and psychic discomfort. Methods already discussed for evaluating intangibles are available for measuring the latter two items.

On the cost side, project resource costs include costs of building, site engineering and servicing, shadow priced as necessary. In addition, existing businesses and residents suffer dislocation costs as a result of redevelopment. For businesses these are likely to take the form of removal expenses and possibly lower profit in cases where the activity is moved to alternative premises, total loss of existing benefits if the activity is forced to close, or temporary loss for businesses which remain *in situ*. Compensation payments have been used here as a minimum measure of dislocation costs (Alexander 1974). For residents who move out, there are removal expenses and perhaps higher living expenses elsewhere, as well as a loss of consumers' surplus in terms of the severance of social contacts and possibly of accessibility to urban service facilities. Contingent valuation methods are probably required to measure the lost surplus. Contingent valuation is also required to determine the other intangible costs in Table 12.6 associated with possible loss of urban character. For example, redevelopment might (at least in the view of some people) impair the architectural harmony and/or heritage of the built environment.

As a footnote, it is worth mentioning that one way around the difficulty of measuring the intangible benefits and costs of redevelopment is to measure only project resource costs. Thus Stone (1963) developed an approach for estimating tangible private and public costs of city development, any differences between schemes in respect of intangibles to be weighted judgementally against cost differences. For schemes expected to yield broadly the same intangible net effects, this approach is one of cost-effectiveness.

12.2.2 Town expansion

Table 12.7 lists the aggregate benefits and costs of town expansion. Benefits comprise the increased value of land taken for development from agricultural or other non-development uses. Government may again appropriate a proportion of these benefits. Costs comprise the capital and continuing resource costs of expansion, shadow priced as appropriate, as well as the intangible costs of lost open space. (For definitions of option and existence values, see Ch. 11.) It is clear that benefits and costs are reversed for an analysis of open-space preservation, and that disaggregation according to different groups is possible for a social analysis.

Applications of conventional CBA to the issue of town expansion are limited in number. Two recent examples, however, illustrate the potential of the approach without capturing all effects identified in Table 12.7. Schwind (1977) looks at three sites of proposed development on the

Table 12.7 Benefits and costs of town expansion

Benefits	Costs
increased land productivity	project resource costs
	lost intangible benefits of open space
	environmental/amenity value
	recreational value
	option value
	existence value

island of Oahu in Hawaii, while Willis (1982) evaluates the benefits of preserving 'green belts' around cities in the UK.

The tangible benefits and costs of expansion schemes are measurable *ex ante* from values of comparable parcels of developed land compared to values in non-development uses, and from engineering estimates. For *ex post* analysis, actual values are available. For the remaining items in Table 12.7, more challenging problems of estimation arise. But despite all their shortcomings, hedonic price, contingent valuation and other methods discussed in Chapter 5 for estimating intangibles are available. For instance, some approximation to the amenity value of open space may be derived from residential property prices using the hedonic price method (e.g. Wabe 1971), and Willis (1982) illustrates use of contingent valuation for measuring the willingness-to-pay for access to, and compensation required for deprivation of, the 'green belt'. Although it generates merely a minimum estimate because it ignores consumers' surplus, one way of measuring lost recreation value is to employ entrance prices to private facilities (e.g. golf courses) as these exist in the 'green belt'. More satisfactorily from the conceptual point of view, Schwind uses the Clawson–Knetsch method (Ch. 11) for estimating the value of open space recreational use.[5] Option value of potential future uses is estimated by Schwind on the assumption that the value per household of preserving a possible future park site is less than or equal to the expected consumers' surplus which would be derived from recreational visits to the park, if it were actually a park (see, for example, Chicchetti & Freeman 1971). Option and existence values, of course, may be estimated in principle by the alternative approach of contingent valuation.

12.3 FINANCIAL ANALYSIS

At the least comprehensive level of analysis is purely financial appraisal of town centre redevelopment and city expansion schemes. While the financial implications of plans are important from the point of view of local government and public sector development companies (not to mention the private developer), it is not easy, as Roberts (1974, p. 131) rightly points out, to defend the exclusive use of a method of evaluation which abstracts from external impacts on parties other than the public sector agencies or the developer involved, and which abstracts from all non-monetary implications of plans (e.g. time spent travelling, noise, landscape impact).[6]

For redevelopment schemes, a simple framework for counting financial benefits and costs from the local authority viewpoint is presented in Table 12.8. This basic framework may be adapted to capture the separate financial viewpoints of different agencies involved in development (Lichfield 1967, Lichfield & Wendt 1969). It may also be adapted to cases

Table 12.8 Town centre redevelopment: local authority benefits and costs

Benefits	Costs
receipts from sale of land to private developers	cost of land acquisition, demolition and site improvement
net additions to local authority tax revenues	net increases in running cost of public services
net increases in income from public facility user charges	cost of relocating (or compensating) displaced occupiers

of town expansion, including the creation of new towns (Lichfield & Wendt 1969).

An alternative type of financial appraisal is simple cost minimization, one version of which – threshold analysis – has been developed for cases of town expansion (Kozlowski & Hughes 1967, Hughes & Kozlowski 1968). Due to topography, utility network systems and other existing land uses, towns encounter physical limitations to their spatial growth. These limitations are the thresholds of urban development which can be overcome only at high costs in terms of capital investment. Alternative patterns of town expansion involving different thresholds are evaluated in terms of minimizing development costs. The method may be used for the case of alternative plans for a single town (for a recent illustration, see Smith 1982) or, at the regional planning level, for choosing towns most suitable for expansion (e.g. Scottish Development Department 1968). Besides the lack of benefit estimation and the narrow focus of mere financial appraisal, another shortcoming is the tendency to ignore running costs in many applications (Willis 1980, p. 200).

12.4 SUMMARY AND CONCLUSIONS

Due to the multiplicity of concerns in town planning – economic, social, political and physical – as well as the dominance in the field of physical planners rather than economists, matrix display methods have been the leading form of analysis conducted along cost–benefit lines. The advantage of these methods is that they attempt to capture by use of physical units of measurement and qualitative assessment, together with monetary measurement where possible, all intangible as well as tangible impacts of land use plans on all groups affected. While comprehensive in this sense, they have tended to obscure the aggregate economic effect of plans.

Conventional CBA, by contrast, attempts to measure all effects in monetary terms and to highlight the aggregate economic impact. In principle, this method may also be used to display distribution effects. In practice, measurement of all impacts in monetary terms proves to be a daunting task so that studies have generally been limited in scope and use of the method itself has not been widespread. Nonetheless, relevant models and related empirical procedures are available, awaiting future refinement. There remains, however, some resistance among physical planners concerned about the breadth and reliability of the analysis so far as town planning is concerned (e.g. Self 1975).

At the least comprehensive level of analysis is financial appraisal, which excludes non-monetary impacts of plans altogether and focuses typically on merely the interest of agencies directly involved in the development effort. While limited in scope, this is not to say that pure financial analysis is not worthwhile, since the financial implications of land-use plans are important for the different parties affected.

NOTES

1 It should be noted that differential weighting may be introduced into the PBS; indeed, Lichfield and associates have experimented in some studies with alternative weights (e.g. Lichfield & Chapman 1968, Lichfield 1969).
2 An exception is the city centre redevelopment analysis by Alexander 1974.
3 The Kirk & Sloyan study is well worth reading as a good example of how not to undertake a conventional CBA of a redevelopment scheme. For a catalogue of its divergences from the principles outlined in Part I of this book, see Heald 1979.
4 In the case of profit opportunities, it is necessary to separate out effects resulting from business newly attracted to the city.
5 It is important to avoid double-counting of recreational benefits which may be included implicitly in amenity values.
6 Although we review financial analysis from the point of view of agencies involved in the development effort, it is, in principle, possible to conduct appraisals from the purely financial points of view of all groups affected by schemes, including users.

13
Local health
and social services

Cost–benefit studies of health programmes have about as long a tradition as studies in transportation (e.g. Fein 1958, Mushkin 1963, Klarman 1965), although they have not been as numerous. Studies of personal social service programmes, on the other hand, have neither the same length nor the weight of tradition. We define this latter category of programmes to comprise services for the elderly, the physically handicapped, the mentally ill, the mentally handicapped, deprived and 'at risk' children, and families with persistent financial, health and/or marital difficulties. In a unitary state, services discussed in this chapter are provided at the regional or, more usually, the municipal level. In a federally organized country, provincial or state governments are also involved.

While economists may not have shown as much interest in these fields as in some others, it is also the case that they have not always been greatly encouraged to do so. Especially in the area of personal social services, but also in the pure health field, there has existed among medical and social work professionals what has been termed 'an historical tradition of economic ignorance and almost economic nihilism' (Knapp 1980, p. 288). In areas so replete with intangible, human concerns, it has sometimes been suggested that techniques of economic analysis are out of place. Not only have economic factors been seen as irrelevant, but attempts to define, much less measure, fairly precise relationships have been viewed as pointless. Moreover, use of economic analysis has been said to betray on the part of the analyst a primitive materialistic ideology which endangers the need to take into account in decision making the essential non-economic concerns of the field. Before outlining the types of studies undertaken in this field and looking in detail at the measurement of benefits and costs, an attempt is made to answer these charges.

The argument that economic considerations are irrelevant attaches an implicit weight of zero to economic factors. In situations governed inevitably by resource constraints, however, decisions must necessarily be based on consideration of economic as well as non-economic factors. If a proposed programme is expected to yield net non-economic benefits, for example, it is necessary to ask whether these are worth enough at the margin to warrant the additional expenditure of resources required to secure them. Or would those resources be better used elsewhere? Without some information regarding the opportunity costs of resources to be employed, the judgements required to answer such questions remain less firmly based than they might be.

It is, of course, true that it is seldom possible to define all cost and benefit effects very precisely, much less measure them. But it is another thing to say that attempts to define and measure them are unnecessary. This is the second of the criticisms of the use of economic analysis in health and social services. Its obvious weakness is the fallacy that, since perfection is unattainable, nothing should be tried. The answer to the argument is that it is surely better to have some information regarding what might be measured than to have no information at all. The degree of ignorance concerning a decision area is at least thereby narrowed, if not wholly eliminated. Decision makers must still exercise judgement about the relative importance of measured and unmeasured aspects of the issue and about the quality of some of the measures employed. The analyst has a responsibility here to indicate clearly the basis of his estimates, providing as well an indication of how outcomes may alter as estimates are varied (sensitivity analysis).

The third criticism of the work of the economist is that the use of economic analysis implies that economic considerations alone are of importance. It is perhaps easy to be misled into this belief when, as often as not, it is the purely economic dimensions of an issue which are quantitatively the most tractable and hence those which are often the only ones to be measured. But it should not be taken to imply that the analyst wishes to have *only* economic considerations weighted in the decision-making process. That would be as extremist a view as the one already rejected, that only non-economic considerations are important in the health and welfare field. What the analyst would claim is that by measuring at least some of the dimensions of an issue (in this context, the economic ones) he/she removes the need for decision makers to operate in a complete factual vacuum.

Thus, if a proposal under analysis appears to be worthwhile in terms of tangible measured factors and it is known that it also carries net intangible advantages, then decision makers can feel confident in undertaking it. If, on the other hand, the proposal does not appear to be worthwhile in tangible terms, then decision makers must judge whether or not its net intangible advantages are significant enough to offset the measured net loss. With information on the likely extent of that loss they at least know how much the intangible effects must be reckoned to be worth in order to make the project acceptable. Without such information, they have no such yardstick. Related examples could be mentioned concerning the case of proposals with net intangible effects expected to be negative. In addition, some guidance as to appropriate expenditure adjustments may be given through experimentation with alternative parameter values in the economic model being used.

Although measurement of only some aspects of an issue will not of course provide final answers, it may be said on the above grounds to assist the decision-making effort. It may further be said that there exist possibilities for introducing differential weights on cost and benefit items to reflect relative contribution to intangible objectives, or of specifying constraints to reflect unmeasured concerns. Contingent valuation, hedonic prices, cost effectiveness or matrix display methods may also be used, as we shall see, to deal with intangible considerations.

In summary, then, it seems reasonable to advance the use of economic analysis in the area of health and social services. To suggest that economic concerns are irrelevant is to neglect the existence of resource constraints; while to argue against model building and quantification on the grounds that results are unlikely to be fully precise is to settle for less than it is possible to achieve. Finally, it does not follow that if economic effects alone are measured (as is not necessarily the case anyway), then decisions will ignore the many non-economic concerns so important in the field.

13.1 CLASSIFICATION OF ANALYSES

Following Drummond (1978) and extending his classification beyond merely health care to the provision of personal social services for the needy, we can list four overlapping issues to which studies have been addressed. The first concerns the efficiency of medical procedures for treating illness. An example relates to the relative efficiency of surgical or injection procedures in treating varicose veins (Piachaud & Weddell 1972). From the point of view of urban and regional planning, this category of studies remains of relatively minor interest, its main relevance being related to the improvement of clinical practice. This is not to deny, however, that results point to implications for expenditure decisions within hospital authorities.

A second category of studies is concerned with the timing of treatments. Here the relative advantages of preventive and remedial care are analysed. Thus various screening procedures for early detection of health problems have been evaluated (e.g. screening for Down's Syndrome and spina bifida, Glass 1979); and, as an example from the personal social services, so has the return to preventive social work with 'at risk' children (Schofield 1976a).

A third category of study evaluates alternative locations for treatment. Should those among the needy elderly (Wager 1972) or the mentally ill (Glass 1979, Weisbrod 1981) who could manage with assistance at home be offered care in residential institutions or on a domiciliary basis in the community? Should outpatient services be provided at a central hospital or in peripheral clinics (Glass 1979)? Where should a new outpatient facility be located (Christianson 1976)?

The final category of study concerns the efficiency of treatments for different client groups. There are often difficulties in comparing benefits across client groups with different needs as a result of the problem of measuring health states in commensurable terms. But in cases where comparability is less of a problem – say, between severely subnormal and subnormal cases of mental handicap – the cost–benefit method may be quite useful.

13.2 BENEFITS OF HEALTH AND SOCIAL SERVICE PROGRAMMES

From the point of view of the community as a whole, benefits of health and social services may be divided into resource savings and improvements in patients' or clients' state of health or wellbeing. Treatment or care in the present may be expected to avert later costs of treatment or care, suitably shadow priced. These averted costs include costs incurred by the agency undertaking treatment or care, by other agencies, by friends and relatives who may contribute time to the process of care and who suffer stress and inconvenience in that process, and by patients and clients themselves (in terms of out-of-pocket expenses, lost opportunities for work and leisure, and in pain and suffering). In addition, patients or clients derive better health or living conditions which in turn may enhance earned income opportunities and the enjoyment of leisure time as well as providing an intangible psychic benefit. We may distinguish three approaches to the assessment of benefits: the traditional approach, the willingness-to-pay/compensation required approach and non-monetary evaluation.

13.2.1 Traditional approach

In pioneering studies of health care (e.g. Mushkin 1962, Klarman 1965), benefits are divided into direct (savings in medical resources from averted or corrected illness) and indirect (averted lost production) components. It will be noted that additional costs as averted for non-medical agencies, patients' families and friends, and patients themselves are overlooked. Following the human capital method, discounted gross protected earnings are used for averted lost production. These are adjusted for life expectancy, labour force participation rates and in some cases for the probability of unemployment, own-consumption and the probability that other workers are displaced or drawn into work through employment of the patient (displacement and vacuum effects).[1] Again, external effects on friends and relatives are typically omitted.

The human capital method contains several shortcomings. For one thing, factor market imperfections throw into question the assumption on which the approach is based that labour is paid the value of its marginal product. Secondly, it poses difficulties for the measurement of the value of averted illness or death for persons not in the workforce. Thus, an elderly life is valued at less than a young life, and steps need to be taken to measure the value of housewives' output. Moreover, the human capital approach disregards the value of health *per se*. Quite apart from its advantages in allowing a person to earn income, good health allows the fuller enjoyment of leisure time and is doubtless valued for its own sake, for the peace of mind which it affords. In other words, enhanced lifetime income bears scant relation to what people would be willing to pay for better health or social service support because of the consumption benefits of health and wellbeing which the human capital method omits.

None of these shortcomings, however, invalidate the approach entirely. The foregoing points suggest that use of the approach provides an understatement of true benefits, which may be sufficient if (a) programmes under analysis appear worthwhile even using this low measure of benefit, or (b) unmeasured benefits can be assumed to be roughly common across alternatives under analysis, or (c) measured results are supplemented by indicators of intangible factors to be considered at the decision-making stage.

While most work using the traditional approach has been in assessing the returns to control or eradication of various diseases (e.g. Klarman 1965, Brooks 1970, Weisbrod 1971), the general methodology has also been applied to the operation of local treatment programmes in an effort to assist local authorities with decisions concerning resource allocation between different client groups. As an example, comparisons have been made of the returns to occupational therapy provided through purpose-built municipal facilities for the mentally handicapped, the mentally ill and the physically handicapped (PA Management Consultants 1972). Although this study may have suffered from the exigencies of time involved in the practice of commercial consulting, it illustrates the traditional approach, modified to take account of some of the factors often omitted in that approach. We outline the study below.

The City of Leicester operated two training and rehabilitation centres, one for the mentally disordered and one for the physically handicapped. Returns to these operations, disaggregated by client groups, were estimated using case histories from the records of each facility and the professional judgement and knowledge of staff employed by the Social Services Department. Measured benefits to the community at large were defined as saved day-care costs and training benefits resulting from attendance. It was recognized that additional intangible benefits having to do with improved wellbeing *per se* also existed (e.g. the psychological benefits to patients of improving or maintaining their self-support, the benefits of social interaction, the psychological benefits to friends and relatives relieved of some responsibility for caring for patients). These latter benefits, however, were not measured.

Day-care benefits provided by the facilities accrued both during attendance when alternative day-care provision was not required, and ultimately in everyday living. Only the former day-care benefits were measured so that excluded day-care benefits, perhaps heroically, were assumed to accrue in proportion to measured benefits across patient groups. Day-care benefits included:

(a) income-earning opportunities for friends and relatives relieved of having to care for patients;

(b) savings in the resource costs of services provided by the local authority or other public agencies (mobile meals, home help, social worker and other professional visits, institutional residential cost) and of equivalent services provided by friends and relatives;

(c) avoided private living costs of being at home during the day.

It is to be noted that, following the human capital method, the value of freed leisure time for friends and relatives was excluded.

In the case of mentally disordered patients, training benefits were measured in terms of incremental gross earnings with a crude adjustment for the probability of relapse designed to understate benefits (in conformity with a guiding principle of the analysis).[2] No adjustment was made for the probability of unemployment on the argument presented earlier (see note 1 below) that the effects of social programmes should be analysed separately from the effects of macro-demand management policies, while the probability of participating in the workforce was gauged from the sample analysed. Displacement and vacuum effects were ignored. In the case of physically handicapped patients, training benefits included the improvement of bathing and housework capability as well as earnings capability measured as above.[3] For persons assessed as having developed the ability to bathe themselves (an important objective in many cases), benefits were measured in terms of averted costs of having district nurses visit patients for the purpose of assisting with bathing. For persons assessed as able to undertake housework (or more housework) as a result of treatment, benefits were measured in terms of averted home help costs. Thus all benefits were defined in terms of avoided resource costs or lost output. Sensitivity analysis was employed regarding certain parameter values where uncertainty prevailed.

Selected benefit–cost ratios for cases of mental disorder are shown in Table 13.1. The analysis compared the annualized costs of construction and annual running costs for the first six years of operation of the training

Table 13.1 Training of the mentally disordered: benefit–cost ratios (1970/1)

| | Discount rate 8% | | Discount rate 15% | |
	a*	b+	a*	b+
severely subnormal	2.30	2.41	1.67	1.69
subnormal	3.66	4.63	2.37	2.56
mentally ill	3.44	4.31	2.14	2.33

*Benefits time horizon: 1980/1.
+Benefits time horizon: retirement age of patients.
Source: PA Management Consultants 1972 (Mental Welfare Exercise), p. 11.

facility against benefits generated from attendance during those six years. Benefits were projected forward to different time horizons and benefits and costs were discounted to present value using discount rates of 8% and 15%.

Even without the undoubtedly important unmeasured intangible benefit of therapy, training and rehabilitation programmes for the mentally disordered appeared to be worthwhile from the point of view of the community as a whole. In terms of the purely economic effects analysed, the return was substantially higher in the subnormal category than in the severely subnormal category, and marginally higher than in the mental illness category. Unless unmeasured intangible considerations would alter rankings, the analysis therefore suggested an order of priority for the allocation of limited resources. It was for decision makers to reflect on the unmeasured factors highlighted in the report; to judge to what extent rankings might be altered by them. Thus, while no claim was made that the analysis could provide cut-and-dried conclusions, it was expected to provide at least some of the input required to inform decisions.

Results of the separate analysis of rehabilitating the physically handicapped were not strictly comparable with results for mental disorders since benefits and costs were measured on an annual basis rather than over chosen time horizons. The suggestion, however, was that the programme barely broke even from the community point of view. This was because few cases left the training centre for outside work and relatively infrequent attendance at the centre gave low day care benefits. The report went on to suggest ways in which the rate of return could be improved, notably by increasing usage of available capacity. Again, although less than comprehensive in the sense of abstracting from intangible benefits, the analysis provided information of a kind which could be useful for decision making.

13.2.2   *Willingness-to-pay/required compensation approach*

The more comprehensive willingness-to-pay/required compensation approach captures in a single measure the combined value of all benefits accruing to beneficiaries. The approach, as indicated in earlier chapters, is implemented through behaviour observation or contingent valuation methods.

As to behaviour observation, we have seen in the context of transportation studies how hedonic prices have been used to estimate the cost of pollution and the value of life and safety. In the current context the method may be used to measure the value of life and perhaps illness avoided. We discussed procedures for estimating the value of life from labour market and consumption data in Chapter 10. We also showed how implications from route, mode and speed choices may be drawn regarding the value of leisure time. In principle, findings from these studies might be used for estimating returns to life-saving treatments and for estimating the value of travelling and waiting time incurred by patients or clients, as well as friends and relatives, involved in programmes of treatment. We also discussed in Chapter 10 a procedure for deriving prices for the value of life from observed decisions, namely the values implicit in public expenditure and legislative decisions to promote safety. As an example, the value of life has been estimated (at around £100 000 in 1969) from legislation to enforce the addition (at extra cost) of safety cabs on tractors (Sinclair 1969). Finally, we have emphasized earlier the shortcomings of the method, notably that a wide variety of results is derived from it and that control for extraneous influences on decisions may be difficult.

As we also pointed out earlier, there has not traditionally been a great deal of interest among economists in the contingent valuation method, but use of the method has recently begun to grow. An advantage of the approach in the context of health and welfare services is that in addition to offering the possibility of measuring the value of life, of time saved and illness avoided, it may be used for measuring the willingness-to-pay for any service or the compensation required to forgo it. Thus in addition to examples of its use for measuring the value of life (e.g. Jones-Lee 1976, Jones-Lee *et al.* 1985), there is a sound recent example of its use for valuing a personal social service (Garbacz & Thayer 1983). The authors estimated the value to recipients of a change in the level of a companion service provided in the USA for elderly clients. Under the scheme, older volunteers provide home help and visitation services to other old people in need in order to avert institutionalization and to promote social integration. Recipients were asked how much they would be willing to pay to prevent a given reduction in service (the equivalent variation) and how much in dollar compensation they would require to put up with the reduction (the compensating variation). The latter value exceeded the former, supposedly due to the operation of the income constraint, and the mean of the two measures exceeded the cost savings to be derived from the reduction in service level. Care was taken in framing questionnaires to avoid biases resulting from survey design; it was also argued on the basis of findings elsewhere that other possible sources of bias were unlikely to cause significant distortion.

It is also conceivable that the questionnaire approach might be used to translate physical indices of health or wellbeing (discussed more fully in the next subsection) into monetary values. Patients or clients could be asked to value health or wellbeing points measured along a physical scale. The advantages of the approach notwithstanding, it is to be remembered that aside from the need to guard against biases relating to strategic behaviour, questionnaire design and interview process, the approach assumes that respondents fully understand the implications of different conditions of health or wellbeing and can convert this understanding into monetary values. In the field of health and welfare this may not always be the case, particularly, for example, in respect of programmes for treating the mentally ill or handicapped.

13.2.3 Non-monetary evaluation

Rather than attempt to place a monetary value on all benefits, the alternatives are to use cost-effectiveness or decision matrix (matrix display) methods as discussed in Chapter 5. So far as cost-effectiveness is concerned, alternative projects are assessed in terms of comparative costs given a common objective, or in terms of comparative physical output given a common budget.[4] If the outputs from each alternative are equivalent, a comparative cost analysis gives as clear a solution as a comparative output study involving equivalent cost. However, that is seldom the case. While surgery and injection sclerotherapy treatments can both be used to correct varicose veins, and while care of the elderly can be provided on either a domiciliary or a residential basis, there are qualitative differences in the type of treatment and care afforded in each case. Thus the comparative cost analyses of these alternatives referred to earlier (Piachaud & Weddell 1972, Wager 1972) leave it to decision makers to weigh the comparative intangible benefits against the measured comparative costs. Definitive solutions are not presented. But a considerable amount of useful comparative information is nonetheless made available. It is to be recognized, of course, that the comparative cost method assumes that the alternatives under analysis are all worth undertaking, that the unmeasured benefits in total outweigh the costs.

Recent examples of the decision matrix approach are given in Glass

Table 13.2 Benefits and costs of bacteriuria screening

	A	B	C
benefit (numbers screened)	960	850	700
cost (per '000 screened) (£)	740	470	180
marginal cost (£)	2.45	1.93	0.26

Source: Glass 1979, p. 107.

(1979) and Weisbrod (1981). Analysing alternative methods (A, B, C) of screening children for bacteriuria, Glass displayed the results shown in Table 13.2. If the value of screening was judged to be at least £2.45 per child, alternative A would be chosen; if it was judged to be worth between £1.93 and £2.45, alternative B would be chosen; if between £0.26 and £1.93, alternative C.

Weisbrod compared the hospital- to the community-based approach for treating the mentally ill. As shown in Table 13.3, benefits were defined in multidimensional terms to include, as well as increased earnings resulting from treatment, improved labour market performance, improved consumer decision-making efficiency (measured by way of insurance expenditures and savings behaviour), and improved health (measured by physical indicators of clinical symptoms, patient satisfaction levels and social involvement). Dollar values were estimated where readily available while non-monetary indicators were used elsewhere; and when no clear information was available a question mark was displayed. The same approach was used on the cost side. The balance of monetary benefits and costs was also shown, with decision makers being left to weigh the monetary balance against the various non-monetary effects taken into account. What this matrix display does is to provide as much useful information as possible and assists decision makers in giving consideration to all factors relevant to their decision.

Before leaving the topic of the non-monetary measurement of benefits, it is worth mentioning a line of research activity which has been directed at the problem of measuring the physical output of health and social programmes (e.g. Culyer *et al.* 1971, Wright 1974). This is not a simple matter when output is in incommensurable multiple dimensions and the desire is to reduce output to, ideally, a single index, perhaps as a preliminary step to monetary measurement or to make possible the use of the cost-effectiveness method, or generally to facilitate the decision-making process. Thus for health programmes, output might be identified in the two dimensions of pain relief and improvement in mobility (Culyer *et al.* 1971). For programmes to assist the elderly, Wright (1974) has suggested that output be measured in terms of programme contributions towards the goals of independence, physical and psychological wellbeing and social integration. In order to combine measures of each of these dimensions into a single cardinal index, it is then necesary to weight each of the measures relatively, a procedure which is unlikely to generate universal agreement given the unavoidable judgements involved. In view of this problem, the search for a single, incontestable index may be thought to be futile in most circumstances. It may not be too difficult, however, to rank states of health or wellbeing ordinally in light of the various goals of programmes; and an ordinal ranking may well be helpful to decision makers who are prepared to work with matrix display results.

Table 13.3 Treatment of the mentally ill: costs and benefits per patient (for 12 months following admission to programme)

	Programme		
	A	B	B − A
Costs			
Costs for which monetary estimates have been made			
1. Direct treatment costs ($)			
Mendota Mental Health Institute (MMHI)			
inpatient	3096	94	−3002
outpatient	42	0	−42
Experimental centre programme	0	4704	4704
Total	3138	4798	1660
2. Indirect treatment costs ($)			
Social Service Agencies			
other hospitals (non-MMHI)	1744	646	−1098
sheltered workshops: Madison Opportunity Center, Inc., and Goodwill Industries	91	870	779
Other community agencies			
Dane County Mental Health Center	55	50	−5
Dane County Social Services	41	25	−16
State Dept of Vocational Rehabilitation	185	209	24
Visiting Nurse Service	0	23	23
State Employment Service	4	3	−1
Private medical providers	22	12	−10
Total	2142	1838	−304
3. Law enforcement costs ($)			
overnights in jail	159	152	−7
court contacts	17	12	−5
probation and parole	189	143	−46
police contacts	44	43	−1
Total	409	350	−59
4. Maintenance costs ($)	1487	1035	−452
5. Family burden costs: lost earnings due to the patient ($)	120	72	−48
Total costs for which monetary estimates have been made ($)	7296	8093	797
Other costs			
6. Other family burden costs			
percentage of families reporting physical illness due to the patient	25%	14%	−11%
percentage of family members experiencing emotional stress due to the patient	48%	25%	23%
7. Burdens on other people (e.g. neighbours, co-workers)	?	?	?

Table 13.3 *contd*   Treatment of the mentally ill: costs and benefits per patient (for 12 months following admission to programme)

	Programme		
	A	B	B − A
8. Illegal activity costs: percentage arrests	1.0%	0.8%	−0.2%
percentage arrests for felony	0.2%	0.2%	0.0%
9. Patient mortality costs (percentage dying during the year)			
suicide	1.5%	1.5%	0%
natural causes	0%	4.6%	4.6%

Benefits

Benefits for which monetary estimates have been made
1. Earnings ($)

	A	B	B − A
from competitive employment	1136	2169	1033
from sheltered workshops	32	195	163
Total	1168	2364	1196

Other benefits
2. Labour market behaviour

	A	B	B − A
days of competitive employment per year	77	127	50
days of sheltered employment per year	10	89	79
percentage of days missed from job	3%	7%	4%
number of beneficial job changes	2	3	1
number of detrimental job changes	2	2	0

3. Improved consumer decision making

	A	B	B − A
insurance expenditures ($)	33	56	23
percentage of group having savings accounts	27%	34%	7%

4. Improved mental health status (details shown separately)

A, hospital programme; B, community programme.
Source: Weisbrod 1981, pp. 540–1.

13.3   COSTS OF HEALTH AND SOCIAL SERVICE PROGRAMMES

From the point of view of the community as a whole, costs represent the value of resources used up in the provision of services. In accounting for the economic costs of providing services it is necessary to recognize that in addition to resources provided by the agency directly responsible for the service (the local health authority, the metropolitan Department of Social Services), resources may also be provided by other agencies in the public sector, by voluntary agencies, by friends and relatives of clients or patients and by clients or patients themselves. Unless a purely financial analysis is being conducted from the viewpoint alone of the sponsoring agency, these external costs are to be included.

Thus in an analysis of the community costs of the preventive effort designed to keep children in 'problem' families out of care and out of the courts (Schofield 1976a), the time invested by social workers and staff in the Social Services Department was supplemented by inclusion of the time expended by the following: management assistants from the local authority Housing Department who operate in part as unqualified social workers dispensing advice and assistance to families with financial/housing problems; health visitors from the Health Department who are involved to the extent that they maintain a watching brief over families with children under school age; home-help services; public and private day nursery, play group and nursery school services; the Probation and After-Care Service (involved in an effort to prevent recidivism); and voluntary agencies such as the National Society for the Prevention of Cruelty to Children. Estimates of time involvement were based on field surveys supplemented by information provided by the professionals engaged in the preventive effort. Valuation was based where available on market values for labour time and day attendance; in the case of voluntary agencies, surrogate market values were chosen.

As another example, in comparing the costs of domiciliary and residential care for the aged, Wager (1972) included differential costs of food, clothing and accommodation incurred by patients themselves or their families. Ideally as well, the time dedicated to patient support by friends and relatives should be costed in terms of willingness-to-pay for freedom from it or compensation required for providing it. Failing this, minimum estimates may be made in terms of forgone income (in so far as people would otherwise have entered the workforce), forgone output of alternative unpaid services (in the case of the housewife) and leisure time. Problems of measuring the values of these different types of time have been discussed in Chapter 10.

In measuring the true resource costs of services it goes without saying that shadow pricing should be considered, not just in respect of intangibles such as the value of time discussed above, but also in cases of market imperfections, unemployment, indirect taxes and transfer payments. That said, it is not clear to what extent costs should be adjusted in all cases. For instance, what is the appropriate degree of adjustment for the effect of doctors' monopoly power on their earnings or for the effect of unemployment among junior nurses or social workers? On the other hand, examples of shadow price adjustments to reflect the opportunity costs of buildings and land used for residential care are found in Weisbrod (1981) and of home nursing (by removing indirect taxes from domiciliary support costs) in Wager (1972). In several studies conducted for the City of Leicester (PA Management Consultants 1972), property taxes and civic contributions to national insurance and superannuation were deducted as transfer payments from accounting costs in estimating resource costs of local authority personal social services.

Two final issues are pertinent to the measurement of costs. The first is the problem of joint costs. The difficulty of apportioning common costs in a hospital as between different treatments is emphasized by Drummond (1981). But the problem arises too in the context of, say, services such as day care and board provided in a residential home for the elderly. Apportionment tends to involve fairly arbitrary decisions. Another issue concerns the importance of distinguishing between average and marginal costs, and between long-run and short-run marginal costs. For decisions regarding expansion or contraction of services it is marginal costs, not always readily available in accounting information, which are relevant. Where capital facilities are extended or closed down, it is long-run as opposed to short-run marginal costs which are relevant. Estimation of hospital cost functions is a fairly well established research interest, most of the work to date being American (e.g. Lave & Lave 1970). Meanwhile, some recent work on operating cost functions in the personal social services (e.g. Knapp 1978, Knapp *et al.* 1979) points to the possibilities of estimating short-run marginal costs in that field. Capital expenditures need to be included, of course, if guidance is to be provided on long-run marginal costs.

13.4 SUMMARY AND CONCLUSIONS

There has not always been enthusiasm among health and social service professionals for the use of CBA in these services. Their objections, however, are capable of being answered; and research has proceeded in terms of assessing the efficiency of service procedures, the optimal timing of treatment or care, the optimal location of facilities and the efficiency of treatment or care for different patient/client groups.

One way of measuring programme benefits is in terms of savings in future resource costs and averted lost income, the traditional approach which, among other shortcomings, abstracts from the consumption benefits of treatment and care. A second, more comprehensive approach involves gauging willingness-to-pay or compensation required by use of behaviour observation or contingent valuation methods. As a third approach, benefits may be evaluated in non-monetary terms using cost-effectiveness or decision matrix (matrix display) methods. For other than purely financial analyses conducted from the point of view of the agency responsible for treatment or care, programme costs are to include costs borne by other agencies, patients/clients themselves and friends and relatives. Issues of shadow pricing, joint costs and the measurement of marginal costs are also to be addressed.

NOTES

1   An argument for not adjusting gross lifetime earnings for the probability of unemployment is that the return to social programmes should not be obscured by the effect of demand management policies (Weisbrod 1962). The argument is that we should consider what patients could earn and produce, not what they might actually earn as affected by unemployment. The case for deducting the patient's own consumption from protected earnings applies only when the analyst is interested in assessing benefits for persons other than the patients themselves.

2   Where sheltered employment was involved, values below market earnings were used.

3   Self-support benefits were not considered to be significant for mental welfare cases.

4   It may be noted that studies which assess the benefits of a single type of treatment or care in terms of the avoided cost of an alternative type are the equivalent of comparative cost studies.

14

Capital investment projects and local economic development

As we have stated in earlier chapters, it is possible that urban renewal, transportation and recreation projects may help to promote the growth of incomes and output in certain areas to the extent that economic activity which would not otherwise occur in these areas is stimulated by such projects. A renewed city centre may attract new business to an urban area, improved transportation links may reduce the transport cost disadvantage of certain regions and thereby stimulate development (see, for example, Kraft *et al.* 1971, Georgi 1973, Ch. 3), while some recreational projects are likely to attract incremental tourist spending to an area. Any benefits by way of these dynamic growth effects are to be seen as additional to the conventional benefits of such projects as outlined in earlier chapters. Moreover, the purpose of other capital investment projects may be to stimulate the growth of income in a particular area. An example is the fashion for mounting 'world fairs' by municipal, provincial and state authorities.

Before discussing the details of analysing growth and development effects, it is stressed that these effects may not, in practice, be significant enough in many cases to require detailed treatment. While the central role of transport costs in classical location theory, for example, suggests the importance of transportation links in regional development, and while improved road systems have been explicitly justified in terms of their expected regional development effects (e.g. HMSO 1963a, 1963b), there have been notes of scepticism raised as to the quantitative importance of these effects. Thus Munro (1969) argues that the causes of Appalachian underdevelopment have little to do with inadequate highway facilities for, by reasonable standards, the highway system in that region is quite adequate. In the UK both Dodgson (1974) and Peaker (1976) present evidence of inconsequential impact on regional employment of major road development schemes.[1] In the field of water resource developments, other authors (e.g. Freeman & Haveman 1970) have argued that more plentiful water or the availability of inland waterway transport is not likely to have done much in the way of stimulating regional development in the USA.

Nonetheless, it remains that local development effects may be worth considering. In what follows we first outline the general requirements of their effective measurement before examining in some detail the issue of estimating local multipliers, as these are relevant for the measurement of economic growth impacts.

14.1 LOCAL DEVELOPMENT BENEFITS

It is emphasized at the outset that precise measurement of local impacts may be very difficult indeed. The effects of some projects may be hard to separate from other influences on local development and may themselves be particularly subtle. Ideally, what is needed is a complex multi-equation model specifying the interrelated roles of all relevant influences together with inter-area linkage effects, the latter to account for interdependencies between different geographical areas; in short, a full-scale multi-area econometric model. The complexity of such a design, as well as the empirical problems associated with its estimation, have caused at least one commentator to wonder, at any rate in the case of transportation effects, whether they are not so intricate 'as to defy representation in any practical sense' (Straszheim 1972, p. 219).

However, the following observations are also to be borne in mind. First, it is not the case that the local development effects of every project are almost or completely impossible to trace. The extent to which a recreational project such as a ski slope or a national park attracts out-of-region visitors and, therefore, tourist income may be gauged relatively easily through questionnaire surveys or hotel records. Secondly, while analysis of the distribution effects of projects on all areas may be the counsel of perfection (Haveman 1976), it is sufficient in many circumstances to focus on only a subset of all areas. This, in fact, is the position adopted in the US Water Resources Council (1970) 'principles and standards' for project evaluation (see Ch. 6). Furthermore, in the event of analysing a project from the single point of view of a local jurisdiction, the impact on other jurisdictions may be ignored anyway, unless there are significant inter-jurisdictional feedback effects. As a third point, it is also important to stress that wholly precise measurement of every conceivable effect is not the *sine qua non* of useful analysis. Absolute precision is not really to be expected, as it is not expected in connection with other analytical techniques; acceptable orders of magnitude are generally sufficient with considerable approximation being tolerable.

What, then, are the considerations to be incorporated into the measurement of local development effects should they be considered to be worth estimating? We may distinguish between primary (or direct) and secondary (or indirect) effects.

14.1.1   Primary development effects

Primary effects comprise local value added as a direct result of the project. This in turn comprises local factor incomes (both labour and non-labour incomes) resulting from incremental spending associated with the project. In the case of urban renewal or transportation projects which attract new economic activity to an area, labour and capital income earned in this activity constitutes the measure of direct benefit; in the case of increased tourist spending resulting from a recreation or other project, incremental spending as it materializes in local factor incomes constitutes the direct benefit.

In measuring primary benefits the following considerations are to be noted. First, it is necessary to adjust direct income associated with projects for the proportion of such income which leaks out of the area in question immediately as a result of (a) tax and other payments to higher jurisdictions, and (b) payments to non-resident factors of production, for example, in respect of labour which commutes into the area for employment. Secondly, it is necessary to measure benefits in light of the counterfactual condition: what would have happened anyway so far as economic development is concerned in the absence of the project. This involves taking into account any effects of projects in displacing through competition some activity which would have otherwise occurred and in reducing transfer payments from higher jurisdictions. Regarding transfers, it is to be recognized that employment creation in an area involves loss of income in the form of transfer payments from higher jurisdictions. Transfer payments saved by the jurisdiction in question, of course, do not enter the calculus, being self-cancelling as a gain to the Treasury and a loss to individual recipients (unless a purely financial analysis from the point of view of the Treasury is being conducted, or differential weights are attached to government and recipient interests).

In *ex ante* analyses, estimates of primary expenditure effects must depend on forecasts. In *ex post* analyses, three complementary approaches are available for the measurement of direct development effects: the questionnaire survey approach, the use of reported records and the use of econometric methods. These may be implemented with respect to income data directly or employment data which may be converted into labour and non-labour income. The questionnaire survey and reported records approaches are self-explanatory. The econometric method involves use of regression analysis to establish the effect of a project on local income (or employment), having standardized for the influence of other factors on local growth and development (e.g. Dodgson 1974, Peaker 1976). The conceptual and empirical difficulties involved in this approach are, of course, legion. The advantage of the method is that in principle it provides the most comprehensive estimate of project impact. It takes automatic account of both the counterfactual condition and secondary as well as primary effects. Estimates of the counterfactual condition are made through coefficients on the non-project variables and the constant term in the fitted equation(s), while secondary effects are subsumed under predicted income or employment change values.[2] Moreover, lags in project impact, if they are relevant, may be estimated as averages or in terms of distributed structures. With questionnaires and reported records, by contrast, separate attention has to be given to counterfactual, secondary impact and lag considerations. On the other hand, the magnitude of a project's impact on area growth, though positive, may not be great enough to be identified using the econometric method so that the other approaches may need to be employed.

14.1.2 Secondary development effects

Secondary development effects refer to increases in local income which are induced by multiplier processes following the creation of value added by the project in question. Such effects comprise increased income in related sectors of the economy through backward and forward production linkages with the project. They also comprise increased income through additional spending induced by project value added. In simple multiplier models, the latter increase results from induced consumption spending alone. But further increases in income may also result from (a) induced investment spending if capital expenditures (public or private) are determined in part by local income levels, or (b) induced export spending if exports are endogenous in the sense of being influenced by outside income levels as these are determined in part by area imports.

It is to be recognized that both linkage and expenditure multiplier effects apply only in so far as full employment of local resources or other supply constraints do not prevail. Where projects influence development in depressed urban areas or regions, secondary effects may assume considerable importance. We outline in the following section the measurement of secondary effects through various types of multiplier estimates. If the comprehensive econometric method of impact estimation is not used, it is necessary to work with separate estimates of multiplier values.

14.2 LOCAL MULTIPLIERS

Local income and output multipliers measure ultimate changes in regional income and output with respect to changes in some autonomous component of local expenditure or direct local income created by projects. Employment multipliers relate changes in total employment to changes in autonomously determined employment. The size of the local multiplier depends directly on the proportion of income spent locally or, inversely, on the proportion of income at each round of spending which leaks out of the local spending stream into savings, taxation, reduced transfer payments or import purchases. Its size, therefore, varies directly with the size of the local area in question since import leakages decline as the size of the area increases.

There are three types of local multipliers which have played a prominent role in urban and regional impact analysis: economic base multipliers, Keynesian multipliers and input–output multipliers. We outline each of these in turn, focusing on their uses and limitations and on issues of their estimation.

14.2.1 Economic base multipliers

The simplest type of multiplier derives its origins from the export or economic base theory of growth, the theory which explains development and decline in terms of the performance of an area's export or economic base. The economic base comprises those sectors of economic activity which serve the export market. The economic base multiplier is defined as the ratio of total economic activity to basic activity.

Thus, if it is assumed that employment is proportional to income, the proportion of income spent locally is equal to non-basic or residentiary employment (NBE) divided by total employment (TE). Then, if it is also assumed that average and marginal propensities to spend locally are equal, the multiplier may be defined as:

$$\frac{1}{1 - (NBE/TE)} = \frac{1}{BE/TE} = \frac{TE}{BE} \tag{14.1}$$

where BE = basic employment, and NBE and TE are as defined above.

In this formulation, the economic base multiplier is an employment multiplier, the usual case. But it is clear that the multiplier could as easily be expressed in income or output terms.[3] Key assumptions underlying the concept as formulated here are that employment is proportional to income (which will not be the case if productivity levels differ as between basic and non-basic sectors) and that average and marginal propensities to spend are equal. This latter assumption, in turn, implies that the propensity to spend is constant with respect to income. Except conceivably as a description of long-run relationships, this is not the usual assumption made in macro-economic analysis.[4] It is usually assumed, for example, that the average propensity to consume falls as income rises so that the average propensity to save increases. Similarly, marginal rates of personal income tax rise as income rises.

Apart from these assumptions, other transparent shortcomings of economic base analysis have been identified (e.g. Isard et al. 1960). First, local or residentiary activity may grow independently of any growth in export activity as a result of autonomous investment (private or public) within the area. Secondly, imports and the multiplier effects of import substitution are ignored. Thirdly, the base multiplier underestimates income multiplier effects in so far as interaction among the basic industries themselves and feedback from the residentiary to the basic sector are precluded. Fourthly, differences between local industries in the extent of inter-industry linkages and hence in the size of industry multipliers are not usually captured, although simple extensions of the method which disaggregate according to industrial sector have been demonstrated (Weiss & Gooding 1968, Garnick 1970). Finally, the distinction between export and residentiary activities is often difficult to draw in practice without recourse to expensive direct surveys, and there is some arbitrariness in the usual combination of judgement and short-cut methods of estimation used in their place. These methods include location quotient and minimum requirement techniques.

The location quotient is defined as (e_{ij}/e_{in}) where e_{ij} is the proportion of employment in area j in industry i and e_{in} is the proportion of national employment in industry i. To the extent that $(e_{ij}/e_{in}) > 1$ it is assumed that industry i serves the export market. Thus $(e_{ij} - e_{in})E_{ij}$ represents export employment, where E_{ij} is employment in industry i in area j. With the minimum requirements technique, export employment in industry i is represented by $(e_{ij} - e_{im})E_{ij}$, where e_{ij} and E_{ij} are as defined before and e_{im} is the proportion of area employment in industry i in the area with the lowest such proportion. This lowest proportion is presumed to be the minimum requirement by any area to satisfy local demands so that all employment above that proportion in other areas is considered to be export employment.

Recognizing the simplifications embedded in the location quotient and minimum requirements techniques (e.g. that area production and demand functions are identical across all areas), Mathur and Rosen (1974) have proposed an alternative econometric method for distinguishing base from residentiary activity. Their proposal is that base employment (activity) depends on employment (activity) in the rest of the nation while purely local employment (activity) is insensitive to levels of activity elsewhere. The average local and non-local employment in each industry can then be estimated, using time-series data, from the following equation fitted for each industry i:

$$E_{ij} = \beta_{0i} + \beta_{li}E_w \qquad (14.2)$$

where E_{ij} = employment in industry i in area j, and E_w = national employment in all industries. Average employment $(\bar{E}_{ij})$ in industry i in area j over the period in question is:

$$\bar{E}_{ij} = \beta_{0i} + \beta_{li}\bar{E}_w \qquad (14.3)$$

and $\beta_{0i}/\bar{E}_{ij}$ and $\beta_{li}\bar{E}_w/\bar{E}_{ij}$ represent respectively the proportion of employment in industry i which is localized (insensitive to changes in outside employment) and the proportion which is non-localized. By application of these proportions to actual industrial employment and by aggregation across industries, total area base and non-base employment is obtained. This method, however, like the previous methods, is really only useful in the context of an essentially closed national economy, for if some national industries export to other countries the area base is underestimated.

14.2.2   Keynesian multipliers

More satisfactory than economic base multipliers are Keynesian multipliers which take account explicitly of leakages from the income stream into savings, taxation, lost transfer payments and imports. We begin with the familiar equilibrium income condition:

$$Y = C + I + G + X - M \qquad (14.4)$$

where Y = area income, C = area consumption expenditure, I = area investment expenditure, G = government expenditure in the area, X = area exports, and M = area imports. The structural form of the model then varies according to taste. One possible specification is as follows. The consumption function may be written:

$$C = \bar{C} + b\,(Y - T) \qquad (14.5)$$

where $\bar{C}$ is exogenously determined consumption spending (spending not determined by area income), T represents personal income taxes and b is the marginal propensity to consume out of after-tax income. The tax function may be written:

$$T = \bar{T} + tY \qquad (14.6)$$

where $\bar{T}$ is the income tax floor and t is the marginal income tax rate, assumed constant (equal to the average rate) for simplicity. The import function may be written:

$$M = \bar{M} + m(Y - T) \qquad (14.7)$$

where $\bar{M}$ is exogenous imports and m is the marginal propensity to import; and investment, government spending and exports may be assumed to be exogenous. Thus:

$$I = \bar{I}, \, G = \bar{G} \quad \text{and} \quad X = \bar{X} \quad\quad (14.8)$$

This simple formulation yields the following reduced form expression for area income:

$$Y = \frac{\bar{C} - (b + m)\bar{T} + \bar{I} + \bar{G} + \bar{X} - \bar{M}}{1 - (b - m)(1 - t)} \quad\quad (14.9)$$

from which first derivatives give income multipliers for autonomous spending $(1/[1 - (b - m)(1-t)])$ and tax changes $(-(b + m)/[1 - (b - m)(1-t)])$. Employment multipliers can be derived from income multipliers (e.g. Greig 1971, Davis 1980), although for purposes of CBA, income or output rather than employment impact is what is of basic interest.

By extending the scope of the model, for instance by endogenizing investment spending and exports, it is possible to introduce respectively induced investment (accelerator) effects – the so-called 'supermultiplier' – and inter-area feedback effects (e.g. Brown et al. 1967, Steele 1969, Black 1981).[5] Feedback effects capture the impact of increased area income and hence imports on the income and imports of other areas and the resultant repercussionary effect on the exports and income level of the area in question. As Black (1981) points out, this effect applies to import leakages on induced investment as well as on the initial increase in area spending. If lags in the operation of the multiplier process are considered to be important, these can also be introduced.

In providing a comprehensive discussion of the Keynesian multiplier, Wilson (1968) adds the point that measurement of the multiplicand is to be undertaken with care since the local income component of the initial exogenous expenditure varies with the type of injection made, given differential leakages into imports, capital income remittances and taxes on different types of initial income created (see also Sinclair & Sutcliffe 1978, 1982). It is also important to recognize that in the case of construction activity, there will be both a 'primary' multiplicand, defined as the local value added component of the initial investment, and a 'secondary' multiplicand, defined as the direct local income associated with on-stream operation of the project (Brownrigg 1971).

It is further important to note that when applying the multiplier to increased area income created as a result of a project, the accuracy of any impact assessment will be enhanced if separate multipliers are estimated for different subgroups of the population involved. This is because different groups display different propensities to consume locally, face differential marginal tax rates and may or may not lose income transfers as they fill the jobs created. Thus Davis (1980) distinguishes between the previously employed, the previously unemployed and in-migrants, computing separate multipliers for each group benefiting from a job creation initiative. Tourism multipliers have also been disaggregated according to type of tourist (Archer & Owen 1971, Brownrigg & Greig 1975, Archer 1976). Similarly, it is often useful to use supplementary input–output data to compute differential industry multipliers (Yannopoulos 1973, Lever 1974, McNicholl 1981).

In the usual absence of regional and subregional income and product accounts, however, estimation of local Keynesian multipliers is not a simple task. Most of the effort has originated in the UK, although a limited and little noticed amount of research has also been conducted in North America (Wilson & Raymond 1973, Moore & Sufrin 1974, Wilson 1975, 1977, Davis 1976, 1980). Two broad approaches to estimation have been employed: the non-survey armchair approach using secondary data and numerous assumptions as to relevant values, and the detailed survey approach.

Archibald (1967) provided one of the first attempts to estimate a multiplier using the non-survey approach. On the basis of plausible judgements regarding values for the marginal propensities to consume and to import, he suggested a minimum value for any region of the UK (of approximately 1.2).[6] Using a similar approach, Brown *et al.* (1967) estimated the multiplier for depressed regions in the UK (at 1.28, or 1.24 for small regions). These authors based estimates of the marginal propensity to import (by industry), the main problem so far as estimation is concerned, on a modified location quotient relating the proportion of national employment in industry i in region j to the proportion of the nation's population in region j. Steele (1969, 1972) used statistics on regional trade flows to estimate import–export linkages between regions and the rest of the world, and available required statistics on direct taxes, savings and transfer payments to further reduce the need for pure guesswork. He estimated separate multipliers for each of the UK regions, with and without interregional feedback effects. Finally, using a variety of secondary data sources, Davis (1976) estimated income multipliers for three regions of British Columbia, Canada.

By contrast with the armchair approach, other researchers have employed intensive survey methods to estimate Keynesian multipliers for particular developments. Thus Greig (1971) estimated local multipliers for a pulp and paper mill in Scotland, while various authors estimated multipliers for the impact of educational institutions on local economies (Brownrigg 1973, Wilson & Raymond 1973, Moore & Sufrin 1974, Manning & Viscek 1977, Wilson 1977). Meanwhile, others have focused on what turned out to be quite low multipliers for tourist spending (Archer & Owen 1971, Brownrigg & Greig 1975, Archer 1976). Using both direct survey and secondary data sources, Davis (1980) estimated local multipliers attaching to jobs created for different types of labour in a remote fishing community of Canada. Finally, using purchase and sales data for individual plants, together with information generated by Archibald (1967) and Brown *et al.* (1967), Lever (1974) estimated separate income multipliers for West Central Scotland according to industrial sector and type of plant (large/small, indigenous/branch).[7]

Table 14.1 displays selected values of Keynesian income multipliers estimated for local or regional economies. Clearly, values vary according to the size and industrial structure of the economy affected, the type of spending injection analysed and assumptions involved in constructing and estimating the multiplier model used. To the extent that it gives some idea of the order of magnitude of estimates developed, the table may suggest likely ranges for general use in impact studies.

14.2.3 *Input–output multipliers*

Use of industrial linkage information by Lever (1974) highlights the fact that the standard approach to the estimation of multipliers disaggregated by industry is through input–output accounts. Input–output models allow us to derive sets of multipliers which recognize that the impact of policy on income (or output or employment) varies according to the sector experiencing the initial expenditure change. Input–output models also make possible estimation of production linkage effects which are not incorporated into Keynesian-type multipliers, except in respect of values added among local suppliers of project inputs at the first round.

It is not necessary to outline in complete detail all the issues involved in constructing input–output tables. This has been expertly undertaken in many other places (e.g. Isard *et al.* 1960, Miernyk 1965, Richardson 1972). But it is useful to give a general outline as background to the discussion

Table 14.1 Keynesian multipliers: selected estimates for local and regional economies

	Income multipliers
Archibald 1967	1.20–1.70
Brown *et al.* 1967	1.24–1.28
Steele 1969 – excluding Scotland*	1.19–1.57
– Scotland*	1.70–1.92
Archer & Owen 1971	1.25
Greig 1971[+]	1.44–1.54
Steele 1972 – for 1964	1.09–1.32
– for 1967	1.24–1.30
Brownrigg 1973[+]	1.45–1.80
Davis 1976	1.08–1.35
Davis 1980	1.04–1.12
McGuire 1983	1.25–1.40

* Range encompasses estimates which include interregional feedback effects.

[+] Range encompasses estimates which include induced private investment or government spending.

Table 14.2 Hypothetical input–output transactions table ($m)

		Processing sector						Final demand				
Inputs[†]	Outputs* →	(1) A	(2) B	(3) C	(4) D	(5) E	(6) F	(7) Gross inventory accumulation(+)	(8) Exports	(9) Government purchases	(10) Gross private capital formation	(11) Households	(12) Total gross output
(1) Industry A		10	15	1	2	5	6	2	5	1	3	14	64
(2) Industry B		5	4	7	1	3	8	1	6	3	4	17	59
(3) Industry C		7	2	8	1	5	3	2	3	1	3	5	40
(4) Industry D		11	1	2	8	6	4	0	0	1	2	4	39
(5) Industry E		4	0	1	14	3	2	1	2	1	3	9	40
(6) Industry F		2	6	7	6	2	6	2	4	2	1	8	46
(7) Gross inventory depletion(−)		1	2	1	0	2	1	0	1	0	0	0	8
(8) Imports		2	1	3	0	3	2	0	0	0	0	2	13
(9) Payments to government		2	3	2	2	1	2	3	2	1	2	12	32
(10) Depreciation allowances		1	2	1	0	1	0	0	0	0	0	0	5
(11) Households		19	23	7	5	9	12	1	0	8	0	1	85
(12) Total gross outlays		64	59	40	39	40	46	12	23	18	18	72	431

* Sales to industries and sectors along the top of the table from the industry listed in each row at the left of the table.
† Purchases from industries and sectors at the left of the table by the industry listed at the top of each column.
Source: Miernyk 1965, p. 9.

of the various multipliers which can be derived from input–output tables.

Three basic steps are involved in generating information for multipliers. The first is to describe the input–output relations in the regional economy in terms of dollar flows, as illustrated in Table 14.2, a hypothetical transactions table.[8] Each row of the matrix shows the distribution of an industry's output to other local industries and to the various categories of final demand. Thus industry A sells $10 million worth of output to itself, $15 million to industry B, and so on. Each column shows industry purchases from other industries and payments for factor services. For example, industry A purchases $5 million worth of input from industry B, $7 million from C, pays $19 million in wages, salaries, dividends, interest and rent to households, and so on. The transactions table represents the accounting framework of input–output analysis.

The second step in the computation of multipliers is to construct from the transactions table the input coefficients table (Table 14.3), the columns of which show the dollar amount of inputs required from each source to produce $1 worth of output in any industry. Thus industry A requires $0.16 worth of input from itself, $0.08 from industry B, and so on. The input coefficients are computed as the proportion of total payments going to each source.

Table 14.3 Input coefficients table: direct purchases per dollar of output

Industry producing		Industry purchasing					
		A	B	C	D	E	F
	A	0.16	0.26	0.03	0.05	0.13	0.13
	B	0.08	0.07	0.18	0.03	0.08	0.18
	C	0.11	0.04	0.21	0.03	0.13	0.07
	D	0.17	0.02	0.05	0.21	0.16	0.09
	E	0.06	0.00	0.03	0.36	0.08	0.04
	F	0.03	0.11	0.18	0.15	0.05	0.13

The third step is to construct the Leontief inverse matrix, or table of direct and indirect requirements per dollar of final demand (Table 14.4). This table shows the dollar production directly and indirectly required *from the industry at the top* as a result of a delivery of $1 worth of output to final demand *by each industry at the .left*. If there is an increase in demand for the output of industry A, there will be, as shown in Table 14.3, direct increases in purchases by A of inputs from B, C, D and so on. In addition, when B sells more of its output to industry A, its demand for inputs from other industries also increases. These indirect, as well as the direct, impacts are summarized in Table 14.4. Thus a $1 increase in demand for output A leads to an increase of $1.38 in the output of A when all effects are counted, an increase of $0.25 in the output of B, $0.28 in the output of C, and so on. The table can be computed tediously by iteration. It is also computed more conveniently as the transposed inverse matrix of Table 14.3 (hence its name).

Table 14.4 Leontief inverse matrix:
direct and indirect requirements
per dollar of final demand

	A	B	C	D	E	F
A	1.38	0.25	0.28	0.41	0.27	0.23
B	0.45	1.21	0.16	0.19	0.12	0.24
C	0.27	0.38	1.38	0.23	0.17	0.39
D	0.35	0.25	0.25	1.53	0.65	0.41
E	0.35	0.26	0.31	0.39	1.28	0.25
F	0.38	0.35	0.22	0.30	0.21	1.32

It is emphasized that difficulties in implementation are not absent in the case of input–output analysis and that the multipliers derived from the tables are only as good as the information on which they are based. These difficulties can be summarized under two headings: fixed coefficients and data collection. In so far as tables are not updated on a frequent basis and thus use coefficients fixed over time, the input–output model assumes away input substitution following changes in relative input prices over time, the effect of technical change as it will alter input proportions over time and the appearance and disappearance of industries over time. The model also includes coefficients fixed in the sense that the implicit production and consumption functions are linear and homogeneous. Thus scale effects in production are assumed away, as is the standard non-linear consumption function. So far as data collection goes, there are two basic methods available: the time-consuming and highly costly detailed survey approach and the approximate non-survey approach. The latter involves, for example, using national coefficients at the area level, and such techniques as the location quotient or minimum requirements method for estimation of area imports and exports.

Bearing in mind the fundamental problems of input–output analysis, the preceding information makes possible the estimation of a variety of multipliers. Three types are illustrated in Table 14.5: the output multiplier, the Type I income multiplier and the Type II income multiplier.[9]

The output multiplier for each sector is the sum of the row of coefficients for each sector in Table 14.4; it measures the change in total output of both final and intermediate goods resulting from a $1 change in final demand for the output of a particular sector. Although this multiplier is useful for indicating the extent of structural interdependence

Table 14.5 Local output and income multipliers

Sector	Output multiplier	Direct income change	Direct and indirect income change	Type I multiplier	Direct, indirect and induced income change	Type II multiplier
A	2.82	0.25	0.63	2.52	1.23	4.92
B	2.37	0.32	0.62	1.94	1.21	3.78
C	2.82	0.18	0.58	3.22	1.13	6.28
D	3.44	0.13	0.61	4.69	1.20	9.23
E	2.84	0.18	0.56	3.11	1.11	6.17
F	2.78	0.20	0.59	2.95	1.16	5.80

between sectors, we are usually more interested for impact analyses in income effects.

The Type I income multiplier is the ratio of the direct and indirect income changes to the direct income change resulting from a unit change in final demand for the output of any sector. The direct income change for each sector is the household row of the input–output table converted to coefficient form (i.e. after endogenizing households in the input coefficient table, Table 14.3). The direct and indirect income changes are represented by the sum of each row entry of the Leontief inverse matrix (Table 14.4), showing direct and indirect requirements per dollar of final demand, multiplied by the household coefficient of the corresponding supplying industry.

The Type II income multiplier is the ratio of the direct, indirect and induced income changes to the direct income change due to a change in final demand. The direct, indirect and induced changes are represented by the household column of an expanded Leontief inverse matrix after inclusion of households in that matrix with the processing sector. The Type II multiplier thus incorporates the effects of secondary rounds of consumer spending (as in the Keynesian multiplier) in addition to direct and indirect inter-industry impacts. It does, however, assume a linear homogeneous consumption function which tends to overstate the income effects of changes in final demand.

Other income multipliers may also be calculated by endogenizing final demand sectors other than households into the processing sector, thus making possible modified income multipliers of the types already discussed. Bourque (1969), for example, endogenized state and local government spending, while Hansen and Tiebout (1963) endogenized all final demands except exports. Public and private induced investment, as addressed in the context of Keynesian multipliers, is thereby incorporated into the multipliers. With inter-area input–output tables (e.g. Miller 1969), moreover, it also becomes possible to endogenize area exports (at least to the extent that they are not international), thus taking account of inter-area feedback effects.

Finally, it is necessary to point out that the high cost, in terms of time and money, of constructing survey-based input–output tables, even of moderate size, has spurred a continuing search for short-cut methods of estimating local, industrially disaggregated multipliers. There have been two broad approaches. The first involves the search for non-survey methods of constructing input–output tables as referred to earlier. These methods, however, have been shown to generate serious errors (Czamanski & Malizia 1969, McMenamin & Haring 1974, Schaffer & Chu 1969); in some cases they remain expensive and time-consuming; and they may not be possible for some areas.

More recently, a second approach has been explored which attempts to estimate multipliers without all the information that is needed in the full input–output matrix. Stevens & Trainer (1976) and Drake (1976) established that the major determinant of an industry's output multiplier is the size of its local purchase coefficients (the proportion of demand for the industry's output satisfied locally) and that small changes in individual coefficients in the inter-industry matrix (processing sector) hardly changed the multipliers at all. One suggestion, therefore, is that reasonable multiplier estimates can be made by determining local purchase coefficients from census data and by using these together with national technical coefficients to estimate the matrix of local area coefficients. Another suggestion (Drake 1976, Burford & Katz 1977, 1981) is to estimate the output multiplier solely from data (in published sources or, better, as supplied by individual firms) on local purchase coefficients. Thus Burford & Katz found that the formula:

$$\hat{u}_j = 1 + \left[\frac{1}{1 - \bar{w}} \right] w_j \qquad (14.10)$$

approximates u_j, the output multiplier of industry j, where w_j, $1 \le j \le n$ represents the proportion of inter-industry expenditures of the jth industry spent within the locality, and $\bar{w}$ is the average column total of the input–output coefficient matrix, or intra-area proportion of purchases for all industries within the area. Income and employment multipliers can also be estimated using the same procedure.

Miernyk (1976), however, makes the point that the differences between input–output and short-cut multipliers for local industries are still large enough to raise doubts about short-cut estimates if a reasonable degree of accuracy is required:

If policymakers are willing to settle for crude estimates, there should be no problem. If they are interested in improving the accuracy of impact statements, though, they should . . . support the collection of data which would improve direct-effect estimates. (p. 50)

The ideal solution to the estimation of reliable industry multipliers, in other words, still resides in the construction of reliable local input–output tables.

14.3 SUMMARY AND CONCLUSIONS

There is a case for estimating the impact on local or regional income of projects which attract incremental economic activity or visitor spending to an area. This local development impact may be divided into (a) primary income effects, and (b) secondary multiplier effects resulting from inter-sectoral production linkages and induced spending.

In *ex post* studies, primary effects may be measured using question-naires, reported records or econometric estimation, the latter approach being the most complete. Three broad approaches have been adopted for estimating multipliers. Of these, the economic base multiplier is the crudest although it does implicitly capture both inter-sectoral linkage and induced spending effects, including in the latter case induced investment and exports as well as consumption spending. The Keynesian multiplier fails to capture the full effect of inter-sectoral production linkages but can be specified to include comprehensive induced spending effects. The input–output approach offers the opportunity to combine both production linkage and induced spending effects while also providing industry-specific multipliers for which the other approaches are less suitable.

Faced with the choice of which type of multiplier to use in practice, the analyst will choose, in light of what is available, what his/her resources permit in the way of own-estimation, the degree of accuracy required and the particular purpose at hand; for example, whether or not it is important to have sector-specific multipliers. Economic base, armchair Keynesian and short-cut input–output multipliers are the simplest and cheapest to estimate; it goes without saying that they are also likely to be the least accurate. One sensible procedure regarding the use of Keynesian multipliers in situations where the analyst feels unable to generate his own estimates is to employ a range of values (sensitivity analysis) drawn from available estimates for areas similar in size and economic structure to the one in question. For greater accuracy, detailed Keynesian income and product accounts or input–output accounts are required, depending on whether Keynesian or input–output-type multipliers are chosen.

NOTES

1 This is not to say, however, that the impact on specific municipalities or other subregional areas might not be significant.
2 A caveat needs to be entered concerning use of the constant term in defining

the counterfactual situation. This is that the method assumes that estimated relationships between dependent and independent variables are the same outside the sample population as within it.

3   An example of the estimation of local base multipliers using personal income data is found in Garrison 1972.

4   If data for multiple time periods are available it is possible to estimate a marginal multiplier ($\Delta TE/\Delta BE$) from a regression equation of the following form: $TE = a + bBE + u$ where a is the constant term, u the stochastic disturbance term and b the multiplier.

5   Alternatively, induced investment (private or public) may be incorporated into impact analysis via the multiplicand (or numerator of the multiplier expression). See Archibald 1967, Greig 1971, Brownrigg 1973, and Davis 1980.

6   A good example of the judgemental approach occurs in the following observation: 'Given the marginal propensity to import (m) for the U.K. as a whole, it seems scarcely possible that m for a region could be less than, say, 0.4' (Archibald 1967, p. 27).

7   For clarification of selected matters involved in Lever's estimation procedure, see McDowall 1975, Swales 1975 and Lever 1975.

8   The numerical examples presented in this section are taken from Miernyk 1965.

9   If the focus of analysis is on the employment-creating effects of projects, employment multipliers can be computed from the input–output table using estimated employment–output relationships for each industry (e.g. Moore and Peterson 1955). An alternative method of estimating employment multipliers is to use employment rather than monetary values as the unit of measurement in building the basic input–output table (e.g. Hansen & Tiebout 1963). However, in CBA the basic interest is in monetary impact.

15
Regional policy evaluation

In recent years the cost–benefit method has been extended to the *ex post* evaluation of policy packages designed to ameliorate the problem of regional disparities in wellbeing. These policies may be classified broadly as either labour supply or labour demand initiatives. Labour supply policies in the regional context comprise programmes to subsidize the movement of surplus labour out of depressed regions to employment elsewhere.[1] Labour demand policies stimulate the employment of labour in depressed regions and include capital investment incentives (grants, tax concessions, loans, loan guarantees), labour employment subsidies, infrastructure improvements to attract new capital spending (including the provision of factory space) and direct controls on the location of new capital development schemes. The two categories of policy are often characterized as respectively 'taking workers to the work' and 'taking work to the workers'.

Evaluation of regional policies may proceed from social, economic or purely financial perspectives, the latter from the point of view of particular firms or individuals relocated under the policy or the Treasury responsible for funding the policy. Examples of labour mobility studies include analyses of selected programmes under the US Manpower and Development Training Act by Black *et al.* (1975), who present both national aggregate economic and relocatees' viewpoints, and Nelson and Tweeten (1973), who merely examine national aggregate economic effects. Among labour demand policies there are US examples of the analysis of local authority capital incentive policies from the viewpoint of the local economy (Rinehart 1963, Sazama 1970, Shaffer & Tweeten 1974, Hellman *et al.* 1976) and the financial viewpoint of the local Treasury (Shaffer & Tweeten 1974). Shaffer & Tweeten further identify the viewpoint of the local private sector and the impact on the national economy. In the UK, Willis (1985) provides examples of attempts to evaluate local authority advance factory construction from the points of view of public sector finances and the national economy. There are also examples of the application of the cost–benefit method to the analysis of broad national regional development programmes of the labour demand type in Canada (ECC 1977, Swan & Glynn 1977, Gillespie & Kerr 1977, Schofield 1978) and the UK (NEDC 1963, Needleman & Scott 1964, Schofield 1976b, Moore & Rhodes 1973, 1974, 1977, Marquand 1980), where varying packages of capital incentives, labour employment subsidies, infrastructure spending and controls on the location of new capital have been used to stimulate job creation in depressed regions. These studies adopt variously the national economic and national Treasury perspectives. Gillespie & Kerr (1977) provide an ambitious social analysis of regional and income class distribution effects of regional development policies.

For two reasons, work in the field of regional policy evaluation, at least regarding large-scale national programmes, remains more tentative than is the case in some of the other fields discussed in earlier chapters. First, there are conceptual complexities associated with the specification of a complete model of the impacts of whole policies compared with single projects. This is especially true for the long term when all manner of structural adjustments may result from policy. Secondly, data requirements are more formidable. Even so, despite the view of some sceptics regarding the value of the so-called 'comprehensive' cost–benefit evaluation of policies (e.g. Marquand 1980, Bartels *et al.* 1982), other authors see the development of an appropriate accounting framework as a matter of urgency (e.g. Holland 1976, Haveman 1976, Armstrong & Taylor 1978, Ch. 11).

In defence of the cost–benefit method it can be reiterated that absolutely precise estimation, while the ideal to be aimed for, is not essential for a technique to have value to decision makers. Approximate, if generally reliable, orders of magnitude may be sufficient. It may be stressed as well that this is really all that is typically achieved with such other techniques of regional and urban analysis as econometric modelling and input–output analysis which have enjoyed a ready acceptance in the field. In this chapter we propose a broad framework in which to analyse the benefits and costs of regional policies before assessing the problems and possibilities of empirical estimation.

15.1 BENEFITS AND COSTS OF REGIONAL POLICIES

In social and economic analyses, benefits and costs should ideally be measured in terms of willingness-to-pay or compensation required for welfare impact in all its dimensions, monetary and non-monetary. However, studies to date have been confined to monetary effects in accordance with a stance adopted by Haveman (1976) in discussing the principles of, and prospects for, regional policy evaluation. This means that measurement of the effect of policies in terms of such intangible considerations as amenity value, impact on individual locational preferences and the like are not given explicit treatment in the interests of empirical convenience. This need not mean, it should be said, that analysis is not useful. Even the narrow monetary focus helps to throw light on policy impact and allows decision makers to weigh the balance between measured and unmeasured effects. If a policy with anticipated unmeasured net advantages, for example, appears to be worthwhile in purely monetary terms, then it can be taken to be worthwhile as well in overall efficiency terms. If a policy with anticipated unmeasured net advantages appears to be less worthwhile in monetary terms than another policy with anticipated unmeasured net disadvantages, then decision makers know by how much the balance of intangible advantages in favour of the first policy is to be valued implicitly if that policy is to be preferred. Limited as the partial framework may be, then, it need not sound the death knell of the cost–benefit method in the context of policy evaluation.

Adopting the monetary focus as described above, the major economic benefits and costs of regional policies from the aggregate point of view of the relevant jurisdiction (nation or region) are listed in Table 15.1. For a social analysis these various benefits and costs may be disaggregated according to population groupings deemed appropriate. Although this is straightforward enough in principle, it could be a challenging empirical task which has been attempted – with respect to regional and income class groupings – in only one known study (Gillespie & Kerr 1977). For a purely financial analysis, components of each type of benefit and cost as respectively appropriated or incurred by the individuals, firms or Treasury in question are separated out.

Table 15.1 Regional policy: aggregate monetary benefits and costs

Benefits	Costs

primary
    incremental value added
	public
	    administrative costs
	    infrastructure costs

secondary
    multiplier effects
	private
	    once-for-all relocation costs
	    continuing locational disadvantage costs

It is clear that the list in Table 15.1 not only excludes the non-monetary implications of regional policies but certain monetary effects as well. So far as non-monetary impacts are concerned, labour mobility policies contribute to growing congestion in the urban centres of prosperous regions, run the danger of violating people's likely preferences for employment in the area where their roots are, contribute to community debilitation in outlying areas and, for these reasons, may also be conducive to political instability. While these implications are not included in the design of Table 15.1, it is possible that they could be grafted on to the framework, either in non-monetary terms using matrix display methods or in monetary terms using contingent valuation or other methods discussed in earlier chapters for measuring intangibles.

So far as monetary effects are concerned, Table 15.1 includes a listing of only what are likely to be the major impacts of policies. Thus possible costs in the way of reduced economic activity in the short and long runs resulting from work disincentive, deadweight loss or interest rate effects associated with the method of financing the policy are excluded. Another example of omission is any long-run loss of economic efficiency resulting from the trade distortion effects of regional incentives as they operate in the manner of non-tariff trade barriers. Reliable approximation of the magnitude of effects such as these remains highly improbable. Yet the assumption of relative insignificance may not be unreasonable.

15.1.1 Primary policy benefits

Primary monetary benefits are identified as incremental value added within the jurisdiction from whose viewpoint the analysis is conducted, that is, policy-generated income accruing to residents of that jurisdiction.[2] Incomes of non-resident factors are excluded as are income deductions such as taxes paid to higher levels of government. In view of the exclusion of non-labour income in most previous studies, it is necessary to stress that income comprises both labour and non-labour income, the latter including that proportion of pre-tax profit and interest, indirect tax and depreciation expenses remaining within the jurisdiction.

A theoretical point underlying this definition of policy benefits concerns the trade-off between inflation and unemployment. In improving the matching of labour market supply and demand by increasing (reducing) demand for labour in areas of excess supply (demand), or by increasing (reducing) supply in areas of excess demand (supply), policy is seen as creating the opportunity to expand the level of economic activity in the economy without fuelling inflation. In other words, the so-called aggregate Phillips curve for the economy as a whole (the trade-off relationship between inflation and unemployment) is shifted leftwards, improving the menu of choice available to society, as shown in Figure 15.1.

The curve to the right in Figure 15.1 shows the combinations of inflation and unemployment available in the absence of policy. The curve to the left shows the combinations made possible through either labour demand or labour supply regional policies, a lower rate of unemployment (higher rate of economic activity) becoming possible at the prevailing inflation rate 'g'. One reason for the shift may be that depressed regions have flatter local Phillips curves than do prosperous regions, so that the moderating effect of policy on inflation in prosperous regions is greater than the reverse effect in depressed regions. Another reason is that policy moderates inflationary wage pressures in prosperous regions as these spread throughout the economy.

The implication is that labour demand policy creates the option of

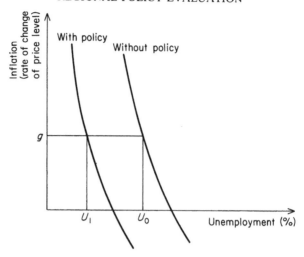

Figure 15.1   Aggregate Phillips curves.

using demand management techniques to replace, in non-assisted areas, jobs diverted from those areas by policy, thus reducing unemployment in the economy as a whole from U_0 to U_1. In the case of labour supply policy, unemployed labour is moved into job vacancies.

The benefit of shifting the Phillips curve leftwards applies even if, in the case of labour demand policy, government recoils in practice from expanding the level of economic activity from U_0 to U_1 because it chooses to observe various constraints on expansion. These may include the state of the balance of payments, the exchange rate, rate of interest, the public sector borrowing requirement or congestion levels in prosperous regions. If this is the case, as Ashcroft (1982) points out, the benefits of regional policy are derived in terms of a reduced inflation rate rather than increased output. But the measure of gain remains the same, the reduction in inflation being implicitly worth the same as the increase in income forgone.

Returning to Table 15.1, definition of incremental value added requires that the counterfactual condition (what would have happened in the way of income creation anyway in the absence of policy) be identified. Here we may distinguish between income from jobs formally attributed to policy but which would have materialized independently without it (autonomous value added) and other income specifically crowded out by policy (displaced value added). The autonomous component of the counterfactual situation includes incremental income which would have occurred in the area irrespective of policy. This element of autonomous income attaches to jobs on which policy subsidies (whether to capital or labour) represent windfall gains, because firms would have located in depressed areas, or labour would have migrated out of depressed areas even without a subsidy (or at least with a smaller subsidy). The autonomous component of the counterfactual situation also includes income that could have materialized, due to autonomous movement, in areas other than those in which policy creates jobs. In the case of labour demand policy, some labour migration out of depressed areas to employment elsewhere is likely to be averted by policy; in the case of labour supply policy, some migration of firms from prosperous to depressed areas in search of ample supplies of labour could also be averted.

Note that to the extent that the Phillips curve hypothesis is accepted, it is not necessary to net out, as an additional element of autonomous income, income attaching to jobs diverted from prosperous to depressed areas under labour demand policy. Similarly, it is not necessary to net out income attaching to jobs diverted to prosperous regions under labour supply policy. In both instances, the reason for not making the further adjustment is that diversion is precisely what makes economic expansion possible.

The second component of the counterfactual condition, policy-displaced value added, comprises in the case of labour demand policy income from jobs eliminated as a result of competition from policy-assisted jobs.[3] It also comprises income associated with intended expansion in prosperous regions precluded by operation of the policy. In this latter connection, it has been pointed out that, prevented under direct location controls from expanding in the area of first choice, some firms may not expand at all (Holmans 1964). In the case of labour mobility policy, displacement may represent the effect of relocation on the productivity of other members of the migrant's family (except to the extent that other family members are replaced in the jobs which they vacate and/or output is otherwise maintained after their departure).

15.1.2 Secondary policy benefits

As in the case of regional development effects stemming from single projects (Ch. 14), secondary income benefits of policy represent the multiplier effects of initial income/expenditure injections. These in turn result from backward or forward production linkages with policy-created jobs and from increased spending induced by initial income increases. Again, secondary benefits occur only to the extent that full employment of resources or other capacity constraints do not prevail. In the case of labour demand policies there can be no question of the relevance of secondary effects, given that policies are designed to combat unemployment in depressed areas. With regard to labour mobility policies, it is necessary to assess the tightness of supply constraints.

15.1.3 Policy costs

The resource costs of policy are identified in Table 15.1 as both public and private. The former comprise costs of administering policy along with the capital cost of infrastructure spending associated with policy, plus continuing expenditures relating to this capital spending. Infrastructure expenditures may be directly earmarked under the policy (e.g. for land preparation and factory construction) or indirectly required to accommodate migrants or to encourage firms to relocate to depressed regions (e.g. improvements in transport facilities, health, education and social services). Several studies to date have excluded consideration of infrastructure costs as part of regional policy; and some of the seminal work in the field (e.g. Needleman & Scott 1964, Moore & Rhodes 1973, 1974, 1977) excludes administration cost as well.

Private costs include the once-and-for-all cost of relocating under the policy on the part of either migrants or firms, and continuing costs of locational disadvantages (e.g. additional living expenses for migrants or additional operating costs for firms which move away from least-cost locations). These private items of cost are designed to be, at least partly, offset by policy subsidies. As explained above, any non-monetary costs of policy, for example additional congestion imposed on prosperous regions by migrants under a labour mobility policy, are not given explicit treatment in the framework under discussion. Part of the intangible costs of congestion, however, are reflected in additional infrastructure spending required to relieve it.

In defining policy costs, it goes without saying that costs are to be shadow priced in accordance with the principles outlined in Chapter 5. It is also necessary to take account of the counterfactual condition as on the benefit side of the analysis. In this connection, deduction of any relocation costs which would have been incurred anyway is required. In the absence of labour mobility policy some migrants would have moved autonomously from depressed regions in search of work and some firms would have moved autonomously to depressed regions in search of labour. Regarding labour demand policy, some firms would have relocated autonomously in the absence of policy (those which receive the policy subsidy as a windfall gain), and some migrants would have otherwise moved from depressed areas in search of work. Any public infrastructure and private costs (once-and-for-all and continuing) which would have been thus incurred in the absence of policy are to be removed for purposes of identifying incremental costs associated with policy. To date studies have paid little attention to counterfactual costs.

15.1.4   *Financial analysis*

From the purely financial point of view of individuals or firms moved under the policy or treasuries of jurisdictions affected by the policy, the foregoing design requires to be adjusted in several respects. In the case of national, state or provincial treasuries, for example, benefits are appropriated out of incremental factor incomes in the form of personal and corporate income tax and social insurance receipts. On the benefit side it is also necessary to take account of reduced unemployment relief payments, increased indirect tax receipts and user charge revenues from facilities (e.g. factories) provided under the programme. At the local government level there is incremental property tax and user charge income. Costs comprise the public policy costs of administration and infrastructure together with policy subsidy payments. If expansionary demand management is used to replace displaced jobs in prosperous areas under labour demand policy, then the costs of government spending outlays and/or reduced tax revenues are also to be included. Another adjustment involves omission of shadow pricing since nominal expenditures are what matter from the financial perspective. Finally, the discount rate used in an economic analysis is replaced by the private rate, the cost of funds to the particular interest being assessed.

15.2   EMPIRICAL ESTIMATION

There can be no doubt that difficulties attend the measurement of the several items to be considered in implementing the framework outlined. The larger the scale of policy and the finer the degree of disaggregation to sub-unit levels of analysis, the greater these difficulties obviously become. But work conducted to date, both within the CBA tradition and elsewhere, suggests that the task is not perhaps impossible, especially with respect to small-scale policies and bearing in mind that round orders of magnitude often suffice. The lessons derived from these sources are assessed in this section.

Three complementary approaches are available for measuring benefits and costs: the questionnaire survey approach, the use of reported records regarding administration of policy and the use of econometric methods. Studies to date have relied mainly on the first two, and particularly on information supplied by sponsoring agencies. The typical procedure has been to measure benefits on the basis of estimates of jobs created by policy, job estimates being developed through either the questionnaire method (e.g. Sazama 1970, Shaffer & Tweeten 1974) or published records (e.g. ECC 1977). To these estimates of job creations are applied measures of income per job, often labour income only (see earlier), and multipliers.[4] Income measures are taken again from either survey information or published data and multiplier values are based on available estimates of Keynesian income multipliers. More completely, multipliers from input–output tables could be employed if available.

So far as the counterfactual position is concerned, it is possible, especially for small-scale programmes such as those administered by local authorities, to conduct research at the firm (or micro) level into the extent of autonomous job creation, job diversion and job displacement. Willis (1980), for instance, reviews examples of the use of (a) studies of the employment generating record of policy-assisted firms relative to non-assisted firms; (b) studies of the labour history of employees; and (c) questionnaire surveys of company officials, as means of estimating autonomous job creation. Another approach is to employ sensitivity analysis involving a range of values for the so-called job incrementality ratio (the ratio of actual to counterfactual jobs) based on a combination of judgement and results from relevant micro studies. Thus, for example, a range of roughly 40%–70% has been used in Canada (ECC 1977) and one of 25%–50% in the UK (Marquand 1980) for labour demand policies. An alternative procedure is not to estimate the incrementality ratio but to demonstrate what minimum value it would need to have in order to make the policy worthwhile (Schofield 1978). Decision makers are then left to exercise their own judgement as to the likelihood that the minimum value is met.[5]

On the cost side, estimates rely on published information regarding public expenditures (with rough prorating for cases of joint costs, e.g. infrastructure spending, administrative cost), and survey results or published subsidy expenditures (as proxies) regarding once-and-for-all and continuing costs of relocation. So far as shadow pricing is concerned, reliance is placed in the few studies where it is included (Schofield 1976b, 1978) on published estimates from earlier work. The alternative is to employ the developing principles outlined in Chapter 5.

As an illustrative example of the approach under discussion, the tentative results of an assessment of the Canadian Federal Government regional industrial development programme are provided in Table 15.2. Deficiencies in data availability compelled preparation of separate cost–benefit estimates according to favourable and unfavourable sets of assumptions. Sensitivity ranges were also used with respect to the shadow pricing of costs, the time horizon of the analysis and the discount rate. As indicated earlier, the value of the incrementality ratio required for the policy to break even was calculated instead of building a specific value for the ratio into estimates. Results, of course, span a wide range and give no clear guidance as to whether the policy was worthwhile or not in aggregate efficiency terms. On the other hand, the purpose of the analysis was to make precisely that point: that firm conclusions as suggested in some commentaries could not be drawn from available data.

Table 15.2 Estimated benefits and costs of
Canadian industrial development programme

		Benefits ($m)	Costs ($m)		Break-even incrementality ratio	
			σ=0.7	σ=1.0	σ=0.7	σ=1.0
Favourable estimates						
5 years	8%	1750.80	18.95	27.04	0.01	0.02
	16%	1401.25	16.91	24.13	0.01	0.02
10 years	8%	4736.83	18.95	27.04	–	0.01
	16%	3246.75	16.91	24.13	0.01	0.01
Unfavourable estimates						
5 years	8%	426.57	314.76	449.60	0.74	1.05
	16%	339.08	281.39	401.94	0.83	1.19
10 years	8%	1605.76	314.76	449.60	0.20	0.28
	16%	1065.17	218.39	401.94	0.26	0.38

σ = Shadow price adjustment factor.
Source: Schofield 1978, p. 26.

Reliance on survey and reported data gives rise to the need to make separate estimates of factors relating to primary, secondary and counterfactual effects. Thus information has to be generated at the micro level regarding the extent to which policy-related developments would have occurred autonomously and the extent to which other developments are displaced by policy. Meanwhile, multiplier estimates have to be generated separately, and such desirable information as input–output relations may or may not be readily available. In addition, it is necessary to make specific estimates of both lags in policy effect (not always revealed in available data) and attrition rates with respect to job creations. Evidence suggests that the latter may often be important in connection with both labour mobility (Nelson & Tweeten 1973, Beaumont 1977) and labour demand policies (DREE 1973, Marquand 1980).

Added to these difficulties is the risk of bias in both data generated by sponsoring agencies and questionnaires. Problems arise concerning exaggeration in estimates of job creation as well as in the definition of counterfactual conditions. Regarding the questionnaire approach, answers by company officials to questions regarding what companies would have done in the absence of policy may well be conditioned by expectations of survey outcome (regarding continuation or modification of the policy) and by the desire to be seen as needing any subsidy received under the policy.

As a way around a number of the above difficulties, the econometric approach has something to recommend it. Only one study in the field has so far used this approach extensively (to estimate overall benefits in terms of incremental investment spending; Hellman *et al.* 1976), although the regression method has been used for estimating selected items of autonomous cost (Schofield 1976b). Nowhere has the method been used to estimate benefits on the basis of job creations. Yet the method has been used widely enough to gauge the employment impact of regional policy in studies of a non-CBA type (e.g. Beacham & Osborne 1970, Sant 1975, Moore & Rhodes 1976, Ashcroft & Taylor 1977, Bowers & Gunawardena 1977). In principle it may be extended to CBA studies.

Ideally, recourse would be had to multi-equation, multiregional models through which the general equilibrium effects of policy could be traced, taking account of interdependencies between regions. Impacts could also be disaggregated by, say, industry or income class, given appropriate specification of models. However, multi-equation models are not likely to be available for many of the jurisdictions under analysis, and rarely do they yet incorporate policy instruments to make possible direct policy assessments. A feasible approach, however, using single equation models would appear to offer a reasonably accessible way of devising estimates for a number of items.

For measuring benefits, the equation takes the following form:

$$IV = f(PI_1 \ldots \ldots PI_n, NI_1 \ldots \ldots NI_n) \qquad (15.1)$$

where IV = impact variable, PI = vector of policy instruments, and NI = vector of non-policy influences on the dependent variable. The dependent variable provides the employment base to which income figures are applied for estimating benefits. It is defined in terms of employment, employment change or geographical moves (of firms or individuals). Policy instruments are measured directly (e.g. the present value of capital incentives) or through dummy variables. Non-policy influences include at least the following: the level of aggregate demand in the economy as a whole (in time-series tests alone) and industrial structure.[6] The equation may be fitted using time-series data over 'policy-on' and 'policy-off' periods. Alternatively, cross-section data for separate sub-areas of a jurisdiction may be used for 'policy-on' periods. The disadvantage of the latter approach is that it precludes estimation of attrition and lag effects.

Time-series use of employment or employment change as the dependent variable leads to the most comprehensive estimate of policy impact. It takes automatic account of the counterfactual condition as well as secondary benefit effects. Estimates of the counterfactual condition are made through the coefficients on the NI variables and the constant term of the fitted equation.[7] Secondary linkage and induced spending effects are subsumed under predicted employment change values. The method also takes automatic account of attrition effects, while lags in policy impact may be estimated either as averages or in terms of distributed structures. Finally, this approach obviates the need to adjust estimates of job creation for possible exaggeration in reported records. The method thus offers considerable convenience over standard approaches.

Time-series use of geographical moves as the dependent variable provides a less comprehensive estimate of policy impact, although it is useful for estimating autonomous effects (through the coefficients on the *NI* variables and the intercept term) and lags. It does not capture displacement or attrition effects and it does not measure the effect of policy in stimulating indigenous expansion (important in connection with labour demand policy). It thus requires greater recourse to supplementary information from surveys or published sources (e.g. input–output tables). On the other hand, it will be more useful than the alternative econometric approach if policy is not powerful enough to be reflected in the aggregate employment changes.

On the cost side, the simple econometric approach may also be useful for estimating certain items. While reported data concerning policy expenditures will still be required and reliance on either assumption or supplementary information will be necessary regarding shadow price adjustments, the method may be employed to gauge the extent to which policy involves incremental infrastructure expenditure. It may also be used in the case of labour demand programmes to estimate the extent of policy-averted labour migration and the level of cost thus avoided.

The econometric approach, then, offers some promise of extending analyses to include a number of items which it may be difficult to estimate by other methods. However, the evidence (Nicol 1982, Ashcroft 1982) is that in measurement of labour demand policy effects, results are markedly sensitive to model specification (variables included, functional form, lag structure) and to variable measures used (e.g. measurement of the pressure of aggregate demand in terms of capacity utilization, vacancies or unemployment). It is also to be reiterated that policy effects may not be powerful enough to be identified using the econometric method.

15.3 SUMMARY AND CONCLUSIONS

Application of the cost–benefit method to the *ex post* analysis of policies which address the problem of regional disparities (policies to encourage labour and capital mobility) remains more experimental than in some of the fields reviewed in earlier chapters. In the case of major national policies, impacts become particularly complex and results correspondingly more approximate and tentative. Indeed, one strand of current opinion (e.g. Marquand 1980, Bartels *et al.* 1982) views analysis of large-scale programmes as being overly ambitious. But even if this is correct, the method remains appropriate for evaluating local development initiatives or small-scale national programmes such as the promotion of labour mobility under the US Manpower and Development Training Act. Emphasis to date has been on measurement of monetary effects, although there is no reason in principle why intangible considerations bearing on welfare should not be included in analyses using matrix display techniques and/or contingent valuation, hedonic prices or market proxy methods.

Empirical estimation has relied largely on piecing together information from surveys and policy records, with little recourse to econometric methods. More extensive econometric estimation, however, offers qualified promise as an additional approach to the measurement of policy impact. As a final remark, it is worth noting that most of the work referred to in this chapter (an exception is Marquand 1980) fails to address the question of whether alternative packages or policies might have been preferable to the single policies analysed. For optimal resource allocation, particular labour demand and labour mobility policies should not only be compared with one another but with alternative packages for achieving the objective of reducing regional disparities; for example, research and development promotion and manpower retraining.

NOTES

1 Manpower training/retraining programmes also serve to improve the matching of vacancies and labour supply and may well assist in addressing the regional problem. However, they are not generally regarded as being exclusively directed at that problem and are not, therefore, included in the discussion of this chapter.

2 It may be noted that in at least one study of a labour demand programme of loans and loan guarantees (Hellman *et al.* 1976), incremental investment spending is used as the measure of economic benefit following conceptual work by Bridges 1965, an approach which fails to reflect the full willingness-to-pay for policy benefits.

3 This category of displacement includes the destructive effect of policy-subsidized operations on competitive enterprises together with the effect of labour poaching from other employment (unless jobs vacated through poaching are filled by the previously unemployed and/or unless output is maintained in original plants through improvement in labour productivity).

4 Some studies omit multipliers, e.g. NEDC 1963, Needleman & Scott 1964, ECC 1977.

5 Still another approach (e.g. Moore & Rhodes 1973) is to assume away the possibilities of (a) autonomous job creation other than through diversion and (b) displacement effects, so that, in the case of labour demand policy, only

averted labour out-migration effects need to be estimated. Econometric methods may be used for this purpose.

6   For use of shift-share analysis as a means of standardizing for the influence of industrial structure in time-series models, see Moore & Rhodes 1976. Willis & Whisker 1980 make the point that the technique breaks down (typically in respect of small areas) when area firms are not representative of firms in the same industries elsewhere in the economy.

7   For a caveat here, see Ch. 14, note 2.

16

CBA in lesser developed countries

The distinctive aspect of the use of CBA in lesser developed countries (LDCs) is that routine procedures make more extensive use of shadow prices than in developed countries. This is because distortions and disequilibria in market processes are more pervasive and serious in LDCs, the distribution of income and wealth is usually more markedly skewed, and the shortage of savings required to finance economic growth is more acute.

Application of the principles of CBA to the realities of LDCs has developed in alternative frameworks outlined in (a) OECD (1969) and Little & Mirlees (1974), and (b) UNIDO (1972). Subsequent work has produced a reconciliation of the two approaches, at least in terms of essentials, such that it is now possible to speak of a unified methodology using the Little & Mirlees (LM) framework as its basis (Squire & van der Tak 1975, Irvin 1978).

In this chapter we outline the unified methodology, discussing issues of shadow pricing before describing a case study by way of illustrating implementation of the method. Shadow price issues include the pricing of traded and non-traded commodities in light of the presence of trade restrictions, efficiency pricing of factors of production and social pricing.

16.1 TRADED AND NON-TRADED COMMODITIES

Suppose that a project produces an export good and uses both domestic and imported inputs.[1] Suppose also that the project is small enough not to alter output or input prices.[2] In the absence of shadow pricing, the official exchange rate (OER) would be applied to the world (or border) price of exports (X) and imported inputs (M), duties and sales taxes being netted out as transfers. Net benefits (NB) of the project would be written:

$$NB = OER(X) - (OER)M - D \qquad (16.1)$$

where D = the value of domestic inputs (for example power, internal transport, labour, land).

As indicated in Chapter 5, however, it is doubtful whether the *OER* accurately reflects the true economic value of foreign exchange to the economy. Given that trade restrictions create an excess demand for foreign exchange at the *OER*, the value of a unit of foreign exchange at the margin, the shadow exchange rate (*SER*), is greater than the *OER*. Thus the domestic value of internationally traded commodities is greater than their border prices expressed in terms of the *OER*.

16.1.1 Shadow pricing

One way to measure the true value of traded commodities is to evaluate them in terms of the *SER* instead of the *OER* (UNIDO 1972) as explained in Chapter 5. The *SER* is estimated by using the ratio of domestic to border prices of traded commodities weighted by the share of each commodity in a country's marginal trade bill. Project net benefits are then written:

$$NB = (SER)X - (SER)M - D \qquad (16.2)$$

or

$$NB = (SER)(X - M) - D \qquad (16.3)$$

Again as explained in Chapter 5, an alternative way of taking account of the true value of traded commodities in a situation involving trade restrictions – the LM method – is to value them at border prices rather than domestic prices, and to shadow price non-traded commodities downwards in order to ensure that all inputs and outputs are valued according to a common yardstick (OECD 1969, Little & Mirlees 1974). Instead of expressing costs and benefits in terms of domestic consumption, everything is expressed in terms of foreign exchange. Thus project net benefits expressed in terms of foreign exchange (*NB'*) are written:

$$NB' = (OER)(X - M) - aD \qquad (16.4)$$

where *a* (the 'standard' conversion factor) = *OER/SER*. Equations 16.3 and 16.4 are clearly equivalent since multiplication of 16.3 by the conversion factor (*a*) yields 16.4.

Hence, apart from details concerning measurement procedures for many elements to be estimated in practice, the difference between the two approaches lies in the choice of yardstick for making traded and non-traded items commensurate. Preference for the LM system is based on its analysis of projects explicitly in terms of net impact on the availability of that scarce resource, foreign exchange. It is also argued that the LM system better reflects the importance of trade efficiency, or the pursuit of comparative advantage, as an objective of economic planning in LDCs. The reason is that border prices represent a set of opportunities open to a country, that is, the actual terms on which it can trade.

The question arises as to how to express non-traded goods at border prices. Non-traded goods are valued by decomposing their marginal costs of production into primary domestic inputs (mainly labour, but also land), traded goods and non-traded goods. Traded goods are then valued at border prices; primary domestic inputs are valued according to procedures outlined in later sections; and non-traded goods are further broken down into the same three constituents – and so on until everything is accounted for in terms of traded goods and primary domestic inputs valued at border prices. In practice, this is usually achieved by the input–output technique. To the extent that it is not possible to disaggregate non-traded goods completed into traded goods and primary inputs, the 'standard' conversion factor (accounting ratio) for the economy as a whole (see Section 16.1.2) may be used to value non-traded goods at border prices.

16.1.2 Accounting ratios

Conversion factors (or accounting ratios) for non-traded commodities and primary domestic inputs are ideally available from the Central Planning Office (and/or the country desk of international aid agencies) for translating domestic prices into border prices. An accounting ratio is the ratio of border price (f.o.b. for exports, c.i.f. for imports) to domestic market price (usually less indirect taxes).[3] Accounting ratios are computed for single commodities, groups of commodities and the whole economy (the 'standard' conversion factor). 'Consumption conversion factors' are also computed for different categories of income (by factor, skill level, region, and so on) as the weighted average ratio for imports and forgone exports in marginal consumption.

A simple example, adapted from Irvin (1978), may help to clarify the procedure. Four commodities are traded. Food is produced for export; machines, textiles and appliances are imported at tariffs respectively of 50%, 25% and 150%. The $OER = \$1 = P1$ where P is the peso. In order to produce $500 (= P500) of food exports, imported machinery worth P300 and labour worth P200 is required; consumption out of wages is food (50%), textiles (25%) and appliances (25%).[4] The issue is whether food production should be increased by $500.

Table 16.1 shows details of the analysis. Without shadow pricing, the project just breaks even (column 1). Using border prices, the project yields a net benefit of $140 (column 2), a result which could be derived by multiplying all costs at domestic prices by their separate accounting ratios (column 3), or by the weighted average conversion factor (accounting ratio) for costs (0.72).

Table 16.1 Increased food production: benefits and costs

	Domestic price (1)	Border price (2)	Accounting ratio((2)/(1)) (3)	Proportion of benefits or costs (4)	Weighted accounting ratio((3)×(4)) (5)
benefits					
food exports	500	500	1.00	1.00	1.00
costs					(0.72)
machines	300	200	0.67	0.60	0.40
labour	(200)			(0.40)	
food	100	100	1.00	0.20	0.20
textiles	50	40	0.80	0.10	0.08
appliances	50	20	0.40	0.10	0.04
net benefit	0	140			

Source: Irvin 1978, pp. 88–9.

Before leaving the topic of valuation in terms of border prices, some final observations are in order. First, traded goods are defined as any input or output that could enter into trade, regardless of whether the commodity is actually imported or exported in the case of the project in question. Secondly, goods which could be traded but which are not currently traded (because of a prohibitive tariff, for example) are treated as traded goods if it is expected that restrictions will be removed, and as non-traded goods otherwise. Thirdly, where there is a difference between the f.o.b. (export) and c.i.f. (import) price of a single commodity, it is convenient to use the mean value. Finally, port handling, inland freight and wholesale distribution charges are to be taken into account as non-traded items in computing the shadow cost of traded commodities.[5] An example concerning the shadow wholesale price of imported bottles in Kenya (Scott *et al.* 1976) is provided in Table 16.2 as an illustration.

Table 16.2 Accounting ratios for imported bottles

Inputs	Proportion of wholesale price (1)	Accounting ratio (2)	Weighted accounting ratio((1)×(2)) (3)
imports c.i.f.	0.592	1.00	0.592
duty*	0.198	0.00	0.000
port handling	0.040	0.77	0.031
rail freight	0.040	0.78	0.031
wholesaling	0.129	0.78	0.101
total	1.000		0.755

* Transfer.
Source: Scott *et al.* 1976, p. 10.

16.2 EFFICIENCY PRICING OF FACTORS OF PRODUCTION

So far we have introduced efficiency prices in respect of trade distortions alone. The separate matter of valuing factor inputs to a project in terms of their opportunity cost to the economy, their forgone value in alternative use, has yet to be addressed.

In perfect markets, factors of production are paid their value in alternative use so that market prices reflect opportunity cost. In LDCs, however, market distortions and disequilibria are prevalent enough that shadow pricing is routinely adopted. In this section, we deal in turn with labour, land and capital.

16.2.1   Efficiency wage rates

For skilled labour, opportunity cost may be adequately reflected in the market wage since markets for skilled labour tend to be reasonably competitive without excessive unemployment. However, where this is not the case, skilled labour is to be shadow priced in the same manner as unskilled labour.

Unskilled labour used as an input to a project in the so-called 'organized' sector is likely to have been drawn from either the informal urban sector or the agricultural sector where substantial unemployment and/or underemployment obtains. In this event, forgone output in the 'unorganized' sector can be expected to be less than the market wage in the organized sector.[6] That the market mechanism, through the process of migration, fails to equalize wage rates in different sectors may be attributed to a variety of factors: skill deficiencies, migration costs, leisure preferences, trade union bargaining power in the 'organized' sector, minimum wage legislation in that sector, and individual assessments of the probability of earning a higher urban wage as balanced against the certainty of earning a given wage in the 'unorganized' sector.

Choice of the measure of forgone output is a matter of judgement. It depends on which segment of the 'unorganized' sector is the source of unskilled labour and on the extent of underemployment in that segment. Some average wage rate/period from the 'unorganized' sector (possibly by region) may be weighted by some measure of the extent of under-employment. In principle, an upward adjustment of this value should also be made (see Ch. 5) to reflect the value of leisure forgone (the disutility of work effort), and any real resource costs (additional living costs, out-of-pocket migration costs, costs of additional infrastructure) associated with movement from the 'unorganized' to the 'organized' sector (Little & Mirlees 1974, p. 172). Given the difficulty of estimating the disutility of work effort, it is commonly assumed for labour drawn from alternative employment that no increased effort (and hence disutility) is involved. In other cases, some estimate of the difference in labour's supply price for new and old jobs is required.[7]

Finally, the estimate of opportunity cost at domestic prices is adjusted by means of an accounting ratio to obtain its value at border prices.

16.2.2   Other efficiency prices

As in the case of labour, it is necessary to express the value of land and capital inputs in terms of forgone alternative value at border prices. In the case of land, market value may be adjusted by the 'standard' conversion factor if land is a relatively unimportant input; otherwise, opportunity cost is multiplied by the accounting ratio for the sector of pre-project employment (Foreign & Commonwealth Office 1972, pp. 12–13).[8]

In the case of capital, its efficiency value is taken into account through the discount rate, termed the accounting rate of interest (ARI) in this context. The ARI is expressed as the marginal product of capital at domestic prices multiplied by the accounting ratio that transfers domestic value into its foreign exchange equivalent. The marginal product of capital is conventionally measured as the social opportunity cost of capital (SOC) discussed in Chapter 8. As indicated in that chapter, alternative proxies are used to approximate its value. Note that the social time preference rate (STP) discussed in Chapter 8 is not included in this context since we deal with situations of non-optimal savings (when $SOC \neq STP$) under a separate heading (Section 16.3).

An alternative approach (for example, Scott et al. 1976, p. 43, Irvin 1978, p. 130) is to treat the ARI as the marginal rate of return on public investment (valued in terms of foreign exchange) rather than the marginal rate of return in the private sector (SOC). In this view, the discount rate is seen as a rationing device whose function is to ensure that the total number of public projects undertaken does not exceed available public saving. When available funds are tight, the ARI is high; when they are plentiful, the ARI is low.

16.3   SOCIAL PRICING

Further to shadow price adjustments for purposes of economic analysis, distributional weights are introduced to compensate for inadequate savings and for inequalities in the distribution of income and wealth. Like efficiency pricing, social pricing has become routine practice in project appraisal in LDCs, despite the objections to social pricing outlined in Chapter 6. Distributional objectives, it is felt, are justifiably addressed through project selection when political considerations and/or weaknesses in tax collection systems preclude the effective use of tax policies for achieving these goals.

The absolute value of differential weights is determined by the numeraire of analysis. In the methodology being followed, the numeraire is uncommitted income in the hands of government (public income) measured in terms of border prices.[9] Uncommitted income refers to income not earmarked for particular purposes. In what follows, we discuss the issue of social pricing separately in its intra- and inter-temporal dimensions, that is, the weighting of consumption benefits respectively between groups (including regions) in the present generation and between generations.

16.3.1 Differential group weights

A factor's social price is its efficiency price adjusted for the social value of increased income accruing to the factor as a result of its employment in the project. The social value of increased factor income depends on (a) the proportion of income consumed, (b) the social value of a unit of additional consumption for group (i) relative to a group at the average level of consumption, and (c) the social value of private sector consumption at the average level relative to the numeraire. This approach highlights consumption out of increased income, recognizing that, while it is a benefit to the factor involved, it is a burden to society if savings are at a premium for purposes of financing investment spending and growth. At the same time, it is recognized that the social worth of additional consumption for different factors depends on their income or wealth class.

Thus, using labour as the example, abstracting from adjustments to the efficiency wage to take account of the disutility of work effort, and from miscellaneous resource costs of employment, and assuming that private savings are as socially valuable as the numeraire, the social wage rate for group (i) (SW_i) may be written:

$$SW_i = \alpha ew_i + c_i \left[(w_i - ew_i) - (w_i - ew_i)d_i/v\right]\beta \qquad (16.5)$$

where ew = efficiency wage rate = forgone output, w = market wage rate, α = accounting ratio for forgone output, c = proportion of increased labour income consumed, β = accounting ratio for increased labour consumption, d = weight reflecting social value of increased consumption for group (i) relative to the 'base' consumption level, and v = value of numeraire relative to private sector consumption at the 'base' consumption level (at which $v = 1$). This expression defines the social wage rate as equal to the efficiency wage rate at border prices plus the net social cost of increased consumption at border prices. The net social cost of increased consumption comprises the extra consumption cost (forgone investment) as a result of employing the worker, and the social assessment of the benefit to the worker of his additional consumption.[10] The net social cost of increased consumption (and hence the social wage rate) is higher, the lower that d_i is (the less important income and wealth redistribution are, and the better off the income recipient is) and the higher v is (the less important consumption relative to the numeraire is). Naturally, judgements form the basis of values for the parameters d_i and v.[11]

The definitional expression for the social price of labour may be readily expanded to take account of forgone leisure (work disutility), additional resource costs of employment and the possibility that $d_i/v \neq 1$ for private sector savings. Also, the social weighting of factor incomes extends in the same manner to increases in income for landowners and capitalists.

16.3.2   Consumption rate of interest

As indicated in Chapter 8, the social time preference rate (STP) has to be given explicit consideration when savings are non-optimal. This is because $SOC \neq STP$. In Chapter 8 we outlined ways of reconciling use of both these relevant rates of discount. In the literature on project appraisal in LDCs, the terminology tends to be different, but the basic principles are the same.

We have seen (Section 16.2) that the ARI corresponds to the SOC, with the qualification that the ARI may also be seen as the marginal rate of return in the public sector. Here we demonstrate that the consumption rate of interest (CRI) corresponds approximately to the STP, recognizing that the CRI, introduced to take account of the value of consumption over time, measures the rate of fall through time in the social value of 'base' consumption.

The weight (W) attaching to a marginal increase in 'base' consumption may be defined as:

$$W = (C_o/C_t)^e \qquad (16.6)$$

where C_o = 'base' consumption today, C_t = 'base' consumption at time t, and e = elasticity of the weight given to marginal increases in 'base' consumption with respect to changes in 'base' consumption. Taking logarithms and differentiating with respect to time, we obtain:

$$\frac{d \log W}{dt} = -e \frac{d \log C_t}{dt} \qquad (16.7)$$

That is, the rate of fall of the weight (the CRI) is equal to the rate of increase in 'base' consumption multiplied by e, or:

$$CRI = ge \qquad (16.8)$$

where g = rate of growth of 'base' consumption. If allowance is now made for 'pure' time preference such that benefits and costs are discounted at STP per cent per annum simply because they occur in the future, then the expression for the CRI becomes:

$$CRI = ge + STP \qquad (16.9)$$

Thus, the CRI equals the STP if the rate of growth of 'base' consumption is zero.

16.4   CASE STUDY ILLUSTRATION

For purposes of illustrating application of the methodology outlined, we look at an appraisal of a highway project in Malaysia (Anand 1976). This is a study that avoids a number of difficult problems of estimation. However, it provides the flavour of what is involved. It also links with the discussion of transportation analysis in Chapter 10.

The most important benefits of the project are savings in vehicle operating costs resulting from a shortened distance between two towns that are the principal generators of traffic for the road. These savings represent consumer surplus for travellers who would have travelled anyway. Other benefits included in the analysis are 'business' time savings, defined as time savings that lead to production of additional marketed goods. These are evaluated at market wage rates of business travellers. Excluded from benefit computations are the following items identified in Chapter 10 as relevant in transport infrastructure appraisals: the consumer surplus gain to newly generated traffic, savings in leisure and commuting time, reduced accident risk and reduced effort. Construction and maintenance costs comprise project costs. All values are expressed in terms of foreign exchange (at 1972 border prices) and the period of analysis is 1972–96.

16.4.1   Standard conversion factor

The 'standard' conversion factor (SCF) is used to convert the domestic value of primary domestic inputs and non-traded goods to foreign exchange values. Ideally, separate conversion factors (or accounting ratios) are required for different categories of goods in question (for example, the consumption of unskilled labour in agriculture or the local output of different construction materials), but ratios were not provided centrally and computation requirements were presumably too demanding. The SCF is calculated from the average import tariff less export duty, with each levy weighted by the appropriate elasticity (demand for imports and supply for exports). On this basis, the *SCF* is estimated at 0.90.

16.4.2 Shadow prices of project inputs

The opportunity cost of rural land taken up by the project is placed at zero in view of the large tracts of virgin land in the area. Urban land is valued at market price and converted to foreign exchange on the basis of the *SCF*.

For purposes of gauging the shadow wage rate, it is necessary to estimate marginal product in alternative employment (the basis of the efficiency wage) as well as the social value of workers' consumption relative to uncommitted public income (the numeraire).[12] The latter value should take into account both the social value (d_i) of increased consumption for group (i) relative to 'base' consumption, and the value (v) of 'base' consumption relative to the numeraire. In this study, estimation of (d_i) is not tackled and (v) is valued at unity. It follows that the social wage rate thus equals the efficiency wage rate. For skilled labour, the market wage is used as the efficiency wage. For unskilled labour, the efficiency wage is estimated as marginal product in rural agriculture plus costs of transfer out of the 'unorganized' sector. All labour costs are converted to border prices by use of the *SCF*.

Besides the primary inputs of land and domestic labour, inputs into construction and road maintenance include foreign and local tools and machinery, foreign and local materials, expatriate labour, overheads and profit (taxes are excluded from each category as transfers). Foreign (traded) inputs are valued directly in terms of foreign exchange. Local (non-traded) inputs are decomposed as far as possible into traded commodities (foreign exchange) and labour, using an input–output table

Table 16.3 Accounting ratios for construction and maintenance

Accounting ratio	Unskilled labour (0.423)*	Skilled labour (0.90)[†]	Foreign exchange (1.0)	Local non-labour (0.90)[†]	Total	Weighted accounting ratio
construction	10.7	7.1	60.9	21.3	100.0	0.910
design and supervision	–	32.0	41.0	27.0	100.0	0.941
routine maintenance	25.0	10.0	48.0	17.0	100.0	0.828
periodic resurfacing	–	10.0	65.0	25.0	100.0	0.965

* The accounting ratio for unskilled labour is the ratio of the efficiency wage to the market wage (0.47) × *SCF* (0.90).

[†] The accounting ratio for skilled labour and non-disaggregated local non-labour inputs is *SCF* (0.90).

Source: Anand 1976, p. 213.

and interviews with local suppliers. Shadow prices for each item of construction or maintenance expenditure are then computed from accounting ratios representing weighted averages of the accounting ratios of its components. Details are shown in Table 16.3.

16.4.3 Shadow prices of project benefits

Savings in vehicle operating costs are measured net of indirect tax (a transfer to government) at border prices. Vehicles, fuel, tyres and engine oil, all traded commodities, are valued directly in terms of foreign exchange.[13] Non-tradable elements of operating costs such as salvage value, crew wages, insurance and vehicle maintenance are converted to border prices by applying the *SCF* to domestic market value.

Inclusion of wages in vehicle operating cost savings takes approximate account of time savings for paid crews. Time saving benefits for passengers of commercial vehicles and passengers and crews of non-commercial vehicles are evaluated separately from consultants' data at market wage rates adjusted to border prices through the *SCF*.

It must be noted, finally, that no differential weighting of benefits according to class of beneficiary is undertaken. The assumption is that income and wealth are optimally distributed among agents of the economy. It is this assumption that allows tax transfers within the economy to be ignored, the social value of an extra dollar of benefit in the hands of any agent, road user or government, being regarded as the same.

16.4.4 Results of the study

Instead of using net present value or benefit–cost ratio criteria, the study evaluates the project in terms of the internal rate of return (*IRR*). For a discussion of the relative merits of these investment criteria, see Chapter 3. The 'best estimate' *IRR* is 17.06%; and sensitivity analysis on key parameters revealed that, under most reasonable assumptions, the return exceeds what is likely to be a minimum return requirement of 10–12% (estimated as either the *ARI* or the *CRI*). The project is, therefore, judged to be economically acceptable.

16.5 SUMMARY AND CONCLUSIONS

Application of CBA in LDCs involves routine use of shadow pricing. This reflects the prevalence of market distortions and disequilibria, maldistribution of income and wealth, and savings shortages. The current chapter has outlined procedures for pricing traded and non-traded commodities in terms of border prices (or foreign exchange) in order to address the effect of trade restrictions in driving a wedge between domestic and border prices. It has also outlined procedures for estimating efficiency prices of factors of production purchased in distorted factor markets, and for estimating social prices designed to take account of redistributional objectives and the social importance of public income and saving relative to consumption.

Finally, a case study was reviewed, illustrating simplified implementation of the methodology outlined. It is clear that detailed implementation requires considerable research effort unless already-estimated accounting ratios are available. Readers interested in seeing how many of the fine details can be addressed in case studies might begin with Little and Scott (1976) and Scott *et al.* (1976).

NOTES

1　This section draws heavily on Irvin 1978, Sect. 4.07.

2　If import or export border prices are altered by the project, marginal import cost and marginal export revenue are substituted respectively for import and export prices.

3　The treatment of indirect taxes (and for that matter also monopoly influences on domestic price) should ideally depend on the principles of shadow pricing outlined in Chapter 5.

4　In order to show how consumption conversion factors are used, this example evaluates the foreign exchange cost of labour's consumption; the cost of project labour also, of course, includes the value of forgone output from alternative employment, and forgone output is to be evaluated in terms of foreign exchange (see Section 16.2.1).

5　A simultaneous equation problem arises here, since in computing accounting ratios for traded goods, it is necessary to know the accounting ratios for traded goods themselves into which the non-traded items are decomposed. Short of computer solutions, the suggested way around this difficulty is initially to estimate accounting ratios for relevant non-traded items using the 'standard' conversion factor. Ratios for traded goods may then be estimated, these ratios may be used to calculate ratios for all non-traded elements, and the latter ratios may be used to recalculate ratios for traded goods (Irvin 1978, p. 96).

6　A possible exception occurs when the creation of a job in the 'organized' sector induces more than one migrant to move from the 'unorganized' sector in search of employment (Mazumdar 1974).

7　Squire and van der Tak (1975, pp. 80–3) emphasize that crude estimates probably suffice in practice.

8　An alternative way to account for the opportunity cost of land is to include producer surplus for the landowner in project benefits. Where this rental element is remitted abroad it is netted out of project benefits.

9　In the UNIDO 1972 approach, the numeraire is aggregate consumption at domestic prices.

10　Define the social value of a marginal increase in consumption for group (i) as:

$$C_i = C_{ci}/C_g$$

where C_{ci} = social value of consumption by group (i), and C_g = social value of uncommitted public income. Let

$$v = C_g/C_{\bar{c}} \text{ and } d_i = C_{ci}/C_{\bar{c}}$$

where $C_{\bar{c}}$ = social value of consumption at the 'base' level of consumption. Thus,

$$C_i = (C_{ci}/C_{\bar{c}})(C_{\bar{c}}/C_g)$$

$$= d_i/v$$

11   Convenient procedures are to set $v = 1$ at the mean level of consumption (Squire & van der Tak 1975) and to reduce the value of d_i in the same proportion as income rises (Scott *et al.* 1976).

12   Anand does not employ an expression for the social wage that is identical to Equation 16.5. The essentials, however, are the same.

13   Since average rather than marginal vehicle operating costs are used, the fixed cost of a vehicle (its new price) is spread over annual mileage and added to variable cost/mile.

17

CBA in urban and regional planning: assessment

Our review has not been intended to cover all conceivable applications of the cost–benefit method at the local planning level because methodologies would be too repetitive. For interested readers there exist examples of the use of CBA for planning the allocation of police resources (Thurow & Rappoport 1969, Blumstein 1971, Shoup 1973, Anderson 1974a), the organisation of refuse collection (IMTA 1969), the level of public transport services (Sugden 1972) and the provision of rural water supplies (Warford & Williams 1971). Examples of other areas of application would be local government fire and education services.

17.1 STRENGTHS AND WEAKNESSES OF CBA

The breadth of applications of CBA indicates the versatility of the technique, as do the various perspectives from which analyses may be conducted: the *ex ante* or *ex post* vantage points in time, different jurisdictional perspectives, aggregate economic and disaggregated social perspectives which go beyond traditional financial analysis. The discussion of applications in Part II of the book also confirms the operational feasibility of the method in many circumstances, even if high degrees of precision are seldom achieved.

It may also be said of CBA, as Drummond (1981) points out, that at the very least it provides a useful framework in which to consider issues. For one thing, it embodies a systematic approach to decision making, a way of thinking methodically about the impacts of decisions rather than flying 'by the seat of the pants'. Secondly, it focuses consideration of these impacts in terms of a consistent body of economic principles. Thus in economic and social analysis, benefits and costs are conceived of in terms of willingness-to-pay or compensation required. In contrast to purely financial analysis, attention is directed to the relevance of external effects, consumer and factor surpluses, intangibles and shadow prices in defining and measuring benefits and costs. The method also emphasizes the importance of identifying the counterfactual condition so that benefits and costs are measured in correct incremental terms.

Of course, the empirical and methodological difficulties associated with the method are not to be underrated. Nor are they to be exaggerated. At the empirical level, deficiencies in data availability and reliability create a host of problems so far as benefit and cost measurement is concerned. We have emphasized the practical difficulties of tracing external and linkage effects; of exactly measuring surpluses; of efficiency pricing when market prices are either inadequate measures of welfare impact or non-existent; of social pricing in regard to distributional considerations; of taking account of risk and uncertainty; and of establishing the minimum return requirement for projects and programmes.

Yet we have also stressed that absolute precision is not the *sine qua non* of useful analysis. The greater the degree of precision, the more useful the analysis obviously is. But a considered attempt to reduce the extent of the factual vacuum in which decisions would otherwise be made can be of considerable value in assisting the decision-making effort. Decision makers will still be called upon to exercise judgement: judgement regarding the reliability of estimates, the relative importance of measured and unmeasured effects, the significance of economic efficiency *vis-à-vis* distributional equity, national security or other social objectives. But partial or approximate information should be better than no information at all, particularly if it is organized according to a body of coherent principles.

It can be said, of course, that partial or approximate estimates of benefit and cost may be more dangerous than helpful. The analyst, however, has a responsibility to make clear the sources of estimates, the assumptions on which they are based and the extent of what he/she has not attempted to measure. On this basis, partial estimates of benefits and costs, minimum or maximum rather than precise values, round orders of magnitude and range estimates can be understood for what they are, as providing limited information. And such information should not only be better than no information at all; it should also be better than spurious precision.

That said, it is important to recognize the many working assumptions required in implementing the principles of CBA. The need for assumptions to make analysis tractable in practice ensures that approximation in estimates is inevitable. Thus, when ordinary, observable demand (and supply) curves are used for the measurement of consumer (and producer) surpluses, we make the implicit assumption that income (or wealth) effects are zero. And, in choosing one of several alternative paths of integration in cases involving related commodities (network effects in the transportation context), we assume that cross-price derivatives are equal. Also, in deciding whether to introduce efficiency prices in the presence of market imperfections, taxes and subsidies, we make assumptions regarding the reaction of the economy to the initiative under analysis; to what extent additional factor supplies are forthcoming as a result of the initiative; to what extent output displaces alternative output.

Furthermore, in adopting contingent valuation, behaviour observation or market proxy approaches to the measurement of intangibles, we accept the several assumptions implicit in each approach. The same point applies in respect of whichever approach is adopted if we introduce social prices to account for distributional considerations; it also applies when we settle on a minimum return requirement and when we choose a procedure for dealing with uncertainty and risk. In addition, in selecting a decision criterion, we accept an implicit assumption concerning reinvestment of annual benefits.

At every step in the analysis, therefore, working assumptions are necessary. However, the need for working assumptions is not unique to CBA. Other techniques in use in urban and regional planning such as input–output analysis, econometric modelling and mathematical programming also involve extensive use of assumptions. The point is not that such use invalidates these techniques but that assumptions should be chosen carefully and, where material, made explicit.

At the methodological level, various reservations attach to the rules and postulates employed in CBA. First, apart from purely financial evaluation, the analysis was traditionally confined to consideration of aggregate efficiency matters. And, due to a lack of confidence in the use of constraints, differential weights and multi-objective matrix displays in some quarters, analysis is frequently still so confined. This restricted focus, however, is not to be interpreted as implying acceptance of the idea that efficiency alone should count. Rather, it indicates scepticism about attempting to do more with CBA than the method is thought to be capable of. It would be a rare economist who would argue that distributional considerations should not be taken into account, at least judgementally, at the decision-making stage.

A second methodological concern relates to the partial equilibrium nature of the cost–benefit method. The method assumes that the general environment in which projects and programmes are implemented is relatively unaffected by implementation. Analysis, therefore, traces some of the ripples created by a decision but certainly not all the general equilibrium effects, the judgement being that beyond readily identifiable effects, others are insignificant. As a result of this judgement, we are able to disregard the problems created by second-best theory and the Scitovsky paradox. It remains, however, the case that disregarded interdependencies in the economy could, in some circumstances, invalidate the piecemeal recommendations based on CBA. It is for this reason that the method is more safely applied, as explained in Chapter 15, to small-scale projects than to larger-scale endeavours such as national regional development programmes.

A third criticism of the methodology of CBA is sometimes heard, which is that it embodies a primitive materialistic ideology in which it is assumed that monetary values can be placed on everything. This is the criticism discussed at the beginning of Chapter 13 regarding the valuation of benefits of health care and social service programmes. As explained there, the criticism is based on the misconceptions that certain qualities are not, in principle, measurable in monetary terms and that when not captured in a monetary measure should not count. In principle, intangibles are measurable in terms of willingness-to-pay or compensation required, although often in practice a complete measure is not achieved. When that is the case, the analyst has a responsibility to alert decision makers to what has not been measured so that these considerations are not overlooked when decision makers come to exercise judgement. Failure to place a monetary value on effects should in no way imply that these effects do not matter. An example is the need to recognize the value of life for its own sake, quite apart from its productive value as measured by discounted lifetime income in many health, transportation and other studies. Moreover, as we have seen, there is scope for the use of non-monetary measures where available. These can be used to supplement the monetary analysis or they can be integrated into cost-effectiveness or matrix display methods.

Another criticism of the methodology of CBA, voiced by Ball (1979) among others, is that it is an ideology based on the assumptions and assertions of neoclassical welfare economics, a point which is emphasized in this book by laying out the welfare economics foundations of the technique in Chapter 2. Ball denies that a social consensus in the form of aggregated preferences can be discerned – through CBA or otherwise – in societies divided by class interest. He further argues that, even if aggregated preferences were relevant, they are not revealed in any objective way through CBA. The assumptions required in aggregating and valuing costs and benefits are variously described as inconsistent, unrealistic and generally adopted merely to facilitate use of the neoclassical paradigm. Non-Marxists, of course, will reject his judgements in favour of their own.

A final concern is that CBA may pre-empt political decision making, replacing it with decision making by technocrats (for example, Wildavsky 1966, Self 1975), or 'econocrats', to use Self's term. The fear is that the technical complexity of CBA may force decision makers to accept the results of analyses unquestioningly instead of attempting to balance the interests of affected parties as expressed through the political process. A competent response to this view is presented by Williams (1973), who takes the position that the analyst has a responsibility to explain the assumptions and procedures used and that nobody should expect a CBA to do more than merely provide one type of input to the decision process. Results, as explained before, are usually sufficiently tentative to be unsuitable as the sole basis for decision making. Taking into account these imperfect estimates together with other relevant considerations, decision makers must ultimately exercise judgement. In this way, CBA can be seen as an aid to, not a replacement of, political decision making.

Like any other method of empirical analysis, then, CBA is seen to have its advantages and limitations. The extent to which the latter are stressed will depend on one's own predilections regarding, for example, the need to reach beyond economic to social analysis, the need for general equilibrium analysis, the morality of placing monetary values on certain intangible considerations, the efficacy of neoclassical economics and the fear of being hoodwinked by 'econocrats'. What has been suggested in the foregoing discussion is that the method is capable of being defended against its critics, and a case made that it can be a useful aid to decision making, if used with care and in full recognition of its limitations so that it is not seen as a panacea for all decision-making problems. Its value in any given circumstance will vary with the complexity of the issue, the availability of reliable data and the competence and honesty of the analyst.

17.2 FUTURE DIRECTIONS

While considerable progress has been achieved in applying CBA to problems of urban and regional planning, scope for further development clearly exists. At one level, there would appear to be scope for further experimentation with promising designs which have not yet been widely applied, and with established designs in areas where they have not yet been widely employed. Examples include methods for estimating net resident benefits in the housing and urban renewal field; use of aggregate economic models – as opposed to matrix display methods – to analyse urban redevelopment and expansion schemes; use of the cost–benefit framework in the area of personal social services; and use of matrix display methods outside the transportation and land use planning fields.

There is also a need to refine methods of measuring intangible impacts. Examples of intangibles which have not been widely measured include costs of dislocation and benefits of reduced social costs and neighbourhood spillovers in connection with urban renewal schemes; benefits of comfort and convenience and costs of dislocation regarding transportation projects; option and existence values together with congestion, ecological damage and social inconvenience costs in respect of recreational projects; congestion and locational preference effects in connection with regional development programmes. In addition, there is scope for refinement of methods used for measuring such other intangibles as the value of life and non-working time, which have received more attention to date.

Of leading importance among methods for measuring intangibles are the hedonic price, time/money cost trade-off and contingent valuation approaches which offer the promise of securing a better handle on several types of intangible effect. Progress is under way in each case, although the need for refinement remains. Further use might also be made of non-monetary measures of impact; for example, in the matrix display format. It might be pointed out, however, that matrix display models themselves could be refined in such a way as to clarify the economic efficiency dimension of issues.

As well as refinement of methodologies for estimating intangible impacts, it would be useful to pursue the goal of establishing standard values (or at least ranges of value) for certain intangibles in order to facilitate everyday implementation of CBA models. It has been seen that this approach has been used in respect of values for non-working time and life. However, estimates of the value of non-working time need to be disaggregated according to different types of time and, as yet, alternative methods have yielded widely differing estimates of the value of life. What this suggests is that not only do methodologies need to be improved but that repeated applications are necessary to build up a store of empirical results from which standard values might be derived with some confidence.

Finally, further refinement of the use of subsidiary techniques for measuring certain tangible items is also required. An example is the use of econometric methods for estimating income or employment impacts of major capital investment projects and regional development policies. We have seen that, at present, estimates are highly sensitive to model specification. Another example is the use of different techniques for estimating local multipliers as used for gauging income or employment impacts. Improvements in these and the other directions discussed above would naturally reinforce the value of CBA in the field of urban and regional planning.

References

Adams, F. G. & Glickman, N. J. (1980), *Modelling the Multiregional Economic System* (Lexington, Mass.: D. C. Heath).

Alexander, I. C. (1974), *City Centre Redevelopment* (London: Pergamon).

Anand, S. (1976), 'The Little–Mirlees appraisal of a highway project in Malaysia', *Journal of Transport Economics and Policy*, vol. 10, no. 3.

Anderson, R. J. & Crocker, T. D. (1971), 'Air pollution and residential property values', *Urban Studies*, vol. 8 no. 3.

Anderson, R. W. (1974a), 'Towards a cost–benefit analysis of police activity', *Public Finance*, vol. 29 no. 1.

Anderson, R. W. (1974b), 'Some issues in recreation economics', in A. J. Culyer (ed.), *Economic Policies and Social Goals* (London: Martin Robertson).

Anderson, R. W. (1975), 'Estimating the recreation benefits from large inland reservoirs', in G. A. C. Searle (ed.), *Recreational Economics and Analysis* (London: Longman).

Archer, B. (1976), 'The anatomy of a multiplier', *Regional Studies*, vol. 10, no. 1.

Archer, B. & Owen, C. B. (1971), 'Towards a tourist multiplier', *Regional Studies*, vol. 5, no. 4.

Archibald, G. C. (1967), 'Regional multiplier effects in the U.K.', *Oxford Economic Papers*, vol. 19, no. 1.

Armstrong, H. & Taylor, J. (1978), *Regional Economic Policy* (Oxford: Philip Allan).

Arrow, K. J. (1963), *Social Choice and Individual Values* (New York: Wiley).

Arrow, K. J. & Lind, R. C. (1970), 'Uncertainty and the evaluation of public investment decisions', *American Economic Review*, vol. 60, no. 2.

Ashcroft, B. (1979), 'The evaluation of regional economic policy: the case of the U.K.', in K. Allen (ed.), *Balanced National Growth* (Lexington, Mass.: D. C. Heath).

Ashcroft, B. (1982), 'The measurement of the impact of regional economic policies in Europe: a survey and critique', *Regional Studies*, vol. 16, no. 4.

Ashcroft, B. & Taylor, J. (1977), 'The movement of manufacturing industry and the effect of regional policy', *Oxford Economic Papers*, vol. 29, no. 1.

Ball, M. (1979), 'Cost–benefit analysis: a critique', in F. Green & P. Nore (eds.), *Issues in Political Economy* (London: Macmillan).

Bartels, C. P. A., Nicol, W. R. & Van Duijn, J. J. (1982), 'Estimating the impact of regional policy: a review of applied research methods', *Regional Science and Urban Economics*, vol. 12, no. 1.

Baumol, W. J. (1968), 'On the social rate of discount', *American Economic Review*, vol. 58, no. 4.

Beacham, A. & Osborne, W. T. (1970), 'The movement of manufacturing industry', *Regional Studies*, vol. 4, no. 1.

Beaumont, P. B. (1977), 'Assessing the performance of assisted labour mobility policy in Britain', *Scottish Journal of Political Economy*, vol. 24, no. 1.

Beesley, M. E. (1965), 'The value of time spent in travelling: some new evidence', *Economica*, vol. 32, no. 126.

Bendavid, A. (1972), *Regional economic analysis for practitioners* (New York: Praeger).

Bergson, A. (1938), 'A reformulation of certain aspects of welfare economics', *Quarterly Journal of Economics*, vol. 52, no. 2.

Black, H. T., Scott, L. C., Smith, L. H. & Sirmon, W. A. (1975), 'On moving the poor: subsidizing relocation', *Industrial Relations*, vol. 14, no. 1.

Black, P. A. (1981), 'Injection leakages, trade repercussions and the regional income multiplier', *Scottish Journal of Political Economy*, vol. 28, no. 3.

Blomquist, G. (1979), 'Value of life saving: implications of consumption activity', *Journal of Political Economy*, vol. 87, no. 3.

Blumstein, A. (1971), 'Cost-effectiveness analysis in the allocating of police resources', in M. G. Kendall (ed.), *Cost–Benefit Analysis* (London: English Universities Press).

Boadway, R. & Flatters, F. (1981), 'The efficiency basis for regional employment policy', *Canadian Journal of Economics*, vol. 14, no. 1.

Bohm, P. (1971), 'An approach to the problem of estimating demand for public goods', *Swedish Journal of Economics*, vol. 73, no. 1.

Bos, H. C. & Koyck, L. M. (1961), 'The appraisal of road construction projects: a practical example', *Review of Economics and Statistics*, vol. 43, no. 1.

Bourque, P. (1969), *The Washington State Economy 1967* (Seattle: Graduate School of Business, University of Washington).

Bowers, J. K. & Gunawardena A. (1977), 'Industrial development certificates and regional policy (Part I)', *Bulletin of Economic Research*, vol. 29, no. 2.

Bridges, B. (1965), 'State and local inducements for industry (Part II)', *National Tax Journal*, vol. 18, no. 2.

Brookes, J. A. & Hughes, K. (1975), 'Housing redevelopment and rehabilitation', *Town Planning Review*, vol. 46, no. 2.

Brooks, R. (1970), 'A cost–benefit analysis of the treatment of rheumatic diseases', *Journal of Economic Studies*, vol. 5, no. 1.

Brookshire, D. & Crocker, T. D. (1981), 'The advantage of contingent valuation methods for benefit–cost analysis', *Public Choice*, vol. 36, no. 2.

Brookshire, D., Thayer, M. A., Schulze, W. D. & d'Arge, R. C. (1982), 'Valuing public goods: a comparison of survey and hedonic approaches', *American Economic Review*, vol. 72, no. 1.

Brookshire, D., Eubanks, L. S. & Randall, A. (1983), 'Estimating option prices and existence values for wildlife resources', *Land Economics*, vol. 59, no. 1.

Brown, A. J., Lind, H. & Bowers, J. (1967), 'The green paper on the development areas', *National Institute Economic Review*, no. 40.

Brown, W. G. & Nawas, F. (1972), 'A new approach to the evaluation of non-priced recreational resources: a reply', *Land Economics*, vol. 48, no. 4.

Brownrigg, N. (1971), 'The regional income multiplier: an attempt to complete the model', *Scottish Journal of Political Economy*, vol. 18, no. 3.

Brownrigg, M. (1973), 'The economic impact of a new university', *Scottish Journal of Political Economy*, vol. 20, no. 2.

Brownrigg, M. & Greig, M. A. (1975), 'Differential multipliers for tourism', *Scottish Journal of Political Economy*, vol. 22, no. 3.

Bruzelius, N. (1979), *The value of time* (London: Croom Helm).

Burford, R. L. & Katz, J. L. (1977), 'Regional input–output multipliers without a full I–O table', *Annals of Regional Science*, vol. 11, no. 3.

Burford, R. L. & Katz, J. L. (1981), 'A method for estimation of input–output type output multipliers when no I–O model exists', *Annals of Regional Science*, vol. 21, no. 2.

Burgess, D. F. (1981), 'The social discount rate for Canada: theory and evidence', *Canadian Public Policy*, vol. 7, no. 3.

Burt, O. R. & Brewer, D. (1971), 'Estimation of net social benefits from outdoor recreation', *Econometrica*, vol. 39, no. 5.

Candler, W. & Boehlje, M. (1971), 'Use of linear programming in capital budgeting with multiple goals', *American Journal of Agricultural Economics*, vol. 53, no. 2.

Case, F. E. (1968), 'Code enforcement in urban renewal', *Urban Studies*, vol. 5, no. 2.

Cassidy, P. A. & Kilminster, J. C. (1975), 'Multiple objectives, public policy and the economist', *Economic Analysis and Policy*, vol. 6, no. 1.

Cesario, F. J. (1976), 'Value of time in recreation benefit studies', *Land Economics*, vol. 52, no. 1.

Cesario, F. J. & Knetsch, J. L. (1970), 'Time bias in recreation benefit estimates', *Water Resources Research*, vol. 6, no. 3.

Cesario, F. J. & Knetsch, J. L. (1976), 'A recreation site demand and benefit estimation model', *Regional Studies*, vol. 10, no. 1.

Cheshire, P. C. & Stabler, M. J. (1976), 'Joint consumption benefits in recreational site "surplus": an empirical estimation', *Regional Studies*, vol. 10, no. 3.

Chicchetti, C. J. & Freeman, A. M. (1971), 'Option demand and consumer surplus: further comments', *Quarterly Journal of Economics*, vol. 85, no. 3.

Chicchetti, C. J. & Smith, V. K. (1973), 'Congestion, quality deterioration and optimal use: wilderness recreation in the Spanish Peaks primitive area', *Social Science Research*, vol. 2, no. 1.

Chicchetti, C. J., Davis, R. K., Hanke, S. H. & Haveman, R. H. (1973), 'Evaluating federal water projects: a critique of proposed standards', *Science*, 24 August, no. 181.

Christianson, J. B. (1976), 'Evaluating locations for outpatient medical care facilities', *Land Economics*, vol. 52, no. 3.

Clawson, M. (1959), *Methods of Measuring Demand for and Value of Outdoor Recreation*, Reprint 10 (Washington, DC: Resources for the Future).

Clawson, M. & Knetsch, J. L. (1966), *Economics of Outdoor Recreation* (Baltimore: Johns Hopkins Press).

Commission on the Third London Airport (1971), *Report, Roskill Commission* (London: HMSO).

Common, M. S. (1973), 'A note on the use of the Clawson method for the evaluation of recreation site benefits', *Regional Studies*, vol. 7, no. 4.

Culyer, A. J., Lavers, R. J. & Williams, A. (1971), 'Social indicators: health', *Social Trends*, no. 2.

Cutt, J. & Tydeman, J. (1976), *A General Approach to the Analysis of Public Resource Allocation* (Canberra: Australian National University Press).

Czamanski, S. & Malizia, E. E. (1969), 'Applicability and limitations in the use of national input–output tables for regional studies', *Papers, Regional Science Association*, vol. 23.

Darling, A. H. (1973), 'Measuring benefits generated by urban water parks', *Land Economics*, vol. 49, no. 1.

Darragh, A. J., Peterson, G. L. & Dwyer, J. F. (1983), 'Travel cost models at the urban scale', *Journal of Leisure Research*, vol. 15, no. 2.

Dasgupta, A. K. & Pearce, D. W. (1972), *Cost–Benefit Analysis* (London: Macmillan).

Davis, H. C. (1976), 'Assessing the impact of a new firm on a small-scale regional economy: an alternative to the economic base model', *Plan Canada*, vol. 16, no. 4.

Davis, H. C. (1980), 'Income and employment multipliers for a small B.C. coastal region', *Canadian Journal of Regional Science*, vol. 3, no. 2.

Davis, O. A. & Whinston, A. B. (1961), 'The economics of urban renewal', *Law and Contemporary Problems*, vol. 27, no. 1.

Davis, O. A. & Whinston, A. B. (1965) 'Welfare economics and the theory of second best', *Review of Economic Studies*, vol. 34, no. 1.

DeBorger, B. L. (1985), 'Benefits and consumption effects of public housing programs in Belgium: some aggregate results', *Urban Studies*, vol. 22, no. 3.

Department of the Environment (UK) (1973), *COBA – A Method of Economic Appraisal for Highway Schemes*, Technical Memorandum H5/73.

Department of Regional Economic Expansion (DREE) (1973), *Assessment of the Regional Development Incentives Program* (Ottawa: DREE).

DeSalvo, J. S. (1971), 'A methodology for evaluating housing programs', *Journal of Regional Science*, vol. 11, no. 2.

DeSalvo, J. S. (1975), 'Benefits and costs of New York City's middle income housing program', *Journal of Political Economy*, vol. 83, no. 4.

Dodgson, J. S. (1974), 'Motorway investment, industrial transport costs and sub-regional growth: a case study of the M62', *Regional Studies*, vol. 8, no. 1.

Doessel, D. P. (1978), 'Economic analysis and end-stage renal disease: an Australian study', *Economic Analysis and Policy*, vol. 8, no. 2.

Doherty, N. & Hicks, B. (1977), 'Cost-effectiveness and alternative health care programs for the elderly', *Health Services Research*, vol. 12, no. 2.

Doling, J. & Gibson, J. G. (1979), 'The demand for new recreation facilities: a Coventry case study', *Regional Studies*, vol. 13, no. 2.

Drake, R. L. (1976), 'A short-cut to estimates of regional input–output multipliers: methodology and evaluation', *International Regional Science Review*, vol. 1, no. 2.

Drummond, M. F. (1978), 'Evaluation and the health service', in A. J. Culyer and K. G. Wright (eds.), *Economic Aspects of Health Services* (Oxford: Martin Robertson).

Drummond, M. F. (1981), 'Welfare economics and cost–benefit analysis in health care', *Scottish Journal of Political Economy*, vol. 28, no. 2.

Dunn, R. M. (1967), 'A problem of bias in benefit–cost analysis: consumer surplus reconsidered', *Southern Economic Journal*, vol. 33, no. 3.

Eckstein, O. (1961a), 'A survey of the theory of public expenditure criteria', in J. M. Buchanan (ed.), *Public Finances: Needs, Sources and Utilization* (Princeton: Princeton University Press).

Eckstein, O. (1961b), *Water Resource Development* (Cambridge, Mass.: Harvard University Press).

Economic Council of Canada (ECC) (1977), *Living Together* (Ottawa: Queen's Printer).

Evans, A. W. (1969), 'Two economic rules for town planning: a critical note', *Urban Studies*, vol. 6, no. 2.

Evans, A. W. (1973), 'Measuring the total impact of a new factory in Furness: a Markovian approach', *Regional Studies*, vol. 7, no. 4.

Fein, R. (1958), *Economics of Mental Illness* (New York: Basic Books).

Feldstein, M. S. (1972), 'The inadequacy of weighted discount rates', in R. Layard (ed.), *Cost–Benefit Analysis* (Harmondsworth: Penguin).

Fisher, A. C. & Krutilla, J. V. (1972), 'Determination of optimal capacity of resource-based recreation facilities', *Natural Resources Journal*, vol. 12, no. 3.

Flegg, A. T. (1976), 'Methodological problems in estimating recreational demand functions and evaluating recreational benefits', *Regional Studies*, vol. 10, no. 3.

Flowerdew, A. D. J. (1971), 'Cost–benefit analysis in evaluating alternative planning policies for Greater London', in M. G. Kendall (ed.), *Cost–Benefit*

Analysis (London: English Universities Press).

Flowerdew, A. D. J. & Rodriguez, F. (1978), 'Benefits to residents from urban renewal: measurement, estimation and results', *Scottish Journal of Political Economy*, vol. 25, no. 3.

Flowerdew, A. D. J. & Stannard, R. (1967), *CBA in Central London Redevelopment*, Urban Studies Conference Paper.

Foreign and Commonwealth Office (1972), *A Guide to Project Appraisal in Developing Countries* (London: HMSO).

Foster, C. D. (1966), 'Social welfare functions in cost–benefit analysis', in R. Lawrence (ed.), *Operational Research in the Social Sciences* (London: Tavistock Publications).

Foster, C. D. & Beesley, M. E. (1963), 'Estimating the social benefits of constructing an underground railway in London', *Journal of the Royal Statistical Society*, Series A, vol. 126, no. 1.

Freeman, A. M. (1970), 'Project design and evaluation with multiple objectives', in R. H. Haveman & J. Margolis (eds.), *Public Expenditure and Policy Analysis* (Chicago: Markham).

Freeman, A. M. (1975), 'A survey of the techniques for measuring the benefits of water quality improvement', in H. M. Peskin & E. P. Seskin (eds.), *Cost–Benefit Analysis and Water Pollution Policy* (Washington, DC: Urban Institute).

Freeman, A. M. (1979), *The Benefits of Environmental Improvement* (Baltimore: Johns Hopkins Press).

Freeman, A. M. & Haveman, R. H. (1970), 'Benefit–cost analysis and multiple objectives: current issues in water resources planning', *Water Resources Research*, vol. 6, no. 6.

Friedman, M. & Savage, L. J. (1948), 'The utility analysis of choices involving risk', *Journal of Political Economy*, vol. 56, no. 4.

Garbacz, C. & Thayer, M. A. (1983), 'An experiment in valuing senior companion program services', *Journal of Human Resources*, vol. 18, no. 1.

Garnick, D. H. (1970), 'Differential regional multiplier models', *Journal of Regional Science*, vol. 10, no. 1.

Garrison, C. B. (1972), 'The impact of new industry: an application of the economic base multiplier to small rural areas', *Land Economics*, vol. 48, no. 4.

Georgi, H. (1973), *Cost–Benefit Analysis and Public Investment in Transport* (London: Butterworth).

Ghosh, D., Lees, D. & Seal, W. (1975), 'Optimal motorway speed and some valuations of time and life', *Manchester School of Economic and Social Studies*, vol. 43, no. 2.

Gibbs, K. C. (1974), 'Recreational land use', in D. W. Pearce (ed.), *The Valuation of Social Cost* (London: Allen & Unwin).

Gibson, J. G. & Anderson, R. W. (1975), 'The estimation of consumers' surplus from a recreation facility with optional tariffs', *Applied Economics*, vol. 7, no. 2.

Gillespie, W. I. & Kerr, R. (1977), 'The impact of federal regional economic expansion policies on the distribution of income in Canada', Discussion Paper no. 895, Economic Council of Canada.

Glass, N. (1979), 'Evaluation of health service developments', in K. Lee (ed.), *Economics and Health Planning* (London: Croom Helm).

Gomez-Ibanez, J. & Fauth, G. R. (1980), 'Downtown auto restraint policies: the costs and benefits for Boston', *Journal of Transport Economics and Policy*, vol. 14, no. 2.

Goodwin, P. B. (1976), 'Human effort and the value of travel time', *Journal of Transport Economics and Policy*, vol. 10, no. 1.

Government of Canada (1976), *Benefit–Cost Analysis Guide* (Ottawa: Treasury Board of Canada).

Greig, M. A. (1971), 'The regional income and employment multiplier effects of a pulp and paper mill', *Scottish Journal of Political Economy*, vol. 18, no. 1.

Hall, P. (1970), *The Theory and Practice of Regional Planning* (London: Pemberton).

Hammack, J. & Brown, G. M. (1974), *Waterfowl and Wetlands: Bioeconomic Analysis* (Washington, DC: Resources for the Future).

Hammerton, M., Jones-Lee, M. W. & Abbott, V. (1982), 'The consistency and coherence of attitudes to physical risk', *Journal of Transport Economics and Policy*, vol. 21, no. 2.

Hansen, W. L. & Tiebout, C. M. (1963), 'An inter-sectoral flows analysis of the California economy', *Review of Economics and Statistics*, vol. 45, no. 4.

Harberger, A. C. (1971), 'Three basic postulates for applied welfare economics', *Journal of Economic Literature*, vol. 9, no. 3.

Harberger, A. C. (1978), 'On the use of distributional weights in social cost–benefit analysis', *Journal of Political Economy*, vol. 86, no. 2.

Harrison, A. J. & Quarmby, D. A. (1969), 'The value of time in transport planning: a review', in *Theoretical and Practical Research on an Estimation of Time-Saving* (Paris and Washington, DC: OECD).

Hartle, D. G. (1976), 'The public servant as adviser: the choice of policy evaluation criteria', *Canadian Public Policy*, vol. 2, no. 3.

Hauer, E. & Greenhough, J. C. (1982), 'A direct method for value of time estimation', *Transportation Research*, vol. 16A, no. 3.

Hausman, J. A. (1981), 'Exact consumer's surplus and deadweight loss', *American Economic Review*, vol. 71, no. 4.

Haveman, R. H. (1965), *Water Resource Investment and the Public Interest* (Nashville, Tenn.: Vanderbilt University).

Haveman, R. H. (1972), *The Economic Performance of Public Investments* (Baltimore: Johns Hopkins Press).

Haveman, R. H. (1976), 'Evaluating the impact of public policies on regional welfare', *Regional Studies*, vol. 10, no. 4.

Haveman, R. H. & Krutilla, J. V. (1968), *Unemployment, Spare Capacity and the Evaluation of Public Expenditures* (Baltimore: Johns Hopkins Press).

Heald, D. (1979), 'Cost–benefit study of the new Covent Garden market: a comment', *Public Administration*, vol. 57, no. 4.

Heggie, I. G. (1979), 'Economics and the road program', *Journal of Transport Economics and Policy*, vol. 13, no. 1.

Hellman, D. A., Wassall, G. H. & Falk, L. H. (1976), *State Financial Incentives to Industry* (Lexington, Mass.: D. C. Heath).

Hensher, D. A. (1972), 'The consumer's choice function: a study of traveller behaviour and values', PhD dissertation, School of Economics, University of New South Wales.

Hensher, D. A. (1976), 'The value of commuter travel time savings', *Journal of Transport Economics and Policy*, vol. 10, no. 2.

Hensher, D. A., McLeod, P. B. & Stanley, J. K. (1974), *Comfort and Convenience: Theoretical, Conceptual and Empirical Extension of Disaggregate Behavioural Models of Modal Choice* (Canberra: Australian Commonwealth Bureau of Roads).

Hensher, D. A. & Truong, T. P. (1985), 'Valuation of travel time savings', *Journal of Transport Economics and Policy*, vol. 19, no. 3.

Hicks, J. R. (1939), 'The foundation of welfare economics', *Economic Journal*, vol. 49, no. 4.

Hicks, J. R. (1943–4), 'The four consumers' surpluses', *Review of Economic Studies*, vol. 11, no. 1.

Hill, M. (1968), 'A goals-achievement matrix for evaluating alternative plans', *Journal of the American Institute of Planners*, vol. 34, no. 1.

HMSO (1963a), *Central Scotland: Programme for Development and Growth*, Cmnd 2188 (London).

HMSO (1963b), *The North East: Programme for Regional Development and Growth*, Cmnd 2206 (London).

Hodge, I. (1982), 'The social opportunity cost of rural labour', *Regional Studies*, vol. 16, no. 2.

Holland, S. (1976), *Capital Versus the Regions* (London: Macmillan).

Holmans, A. E. (1964), 'Restriction of industrial expansion in S.E. England: a reappraisal', *Oxford Economic Papers*, vol. 16, no. 2.

Hughes, J. T. & Kozlowski, J. (1968), 'Threshold analysis – an economic tool for town and regional planning', *Urban Studies*, vol. 5, no. 2.

Institute of Municipal Treasurers and Accountants (IMTA) (1969), *Cost–Benefit Analysis* (London: IMTA).

Irvin, G. (1978), *Modern Cost–Benefit Methods* (London: Macmillan).

Isard, W. et al. (1960), *Methods of Regional Analysis* (Cambridge, Mass.: MIT Press).

Isard, W. et al. (1972), *Ecologic-Economic Analysis for Regional Development* (New York: The Free Press).

Jenkins, G. P. (1973), 'The measurement of rates of return and taxation from private capital in Canada', in W. A. Niskanen et al. (eds.), *Benefit–Cost and Policy Analysis 1972* (Chicago: Aldine).

Jenkins, G. P. & Kuo, C.-Y. (1978), 'On measuring the social opportunity cost of permanent and temporary employment', *Canadian Journal of Economics*, vol. 11, no. 2.

Johnson, F. I. (1985), 'The imprecision of traditional welfare measures in empirical applications', *Applied Economics*, vol. 17, no. 4.

Jones-Lee, M. W. (1976), *The Value of Life: An Economic Analysis* (London: Martin Robertson).

Jones-Lee, M. W., Hammerton, M. & Philips, R. (1985), 'The value of safety: results of a national sample survey', *Economic Journal*, vol. 95, no. 377.

Kaldor, N. (1939), 'Welfare comparisons of economics and interpersonal comparisons of utility', *Economic Journal*, vol. 49, no. 3.

Kirk, J. H. & Sloyan, M. J. (1978), 'Cost–benefit study of the new Covent Garden market', *Public Administration*, vol. 56, no. 1.

Klarman, H. E. (1965), 'Syphilis control programs', in R. Dorfman (ed.), *Measuring Benefits of Government Investments* (Washington, DC: Brookings Institution).

Knapp, M. R. J. (1978), 'Economies of scale in residential care', *International Journal of Social Economics*, vol. 5, no. 2.

Knapp, M. R. J. (1980), 'Planning for balance of care for the elderly: a comment', *Scottish Journal of Political Economy*, vol. 27, no. 3.

Knapp, M. R. J., Curtis, S. & Giziakis, E. (1979), 'Observation and assessment centres for children: a national study of the costs of care', *International Journal of Social Economics*, vol. 6, no. 3.

Knetsch, J. L. & Sinden, J. A. (1982), 'Willingness to pay and compensation demanded: experimental evidence of an unexpected disparity in measures of value', mimeo., Simon Fraser University.

Kozlowski, J. & Hughes, J. T. (1967), 'Urban threshold theory and analysis', *Journal of the Town Planning Institute*, vol. 53, no. 2.

Kraft, G., Meyer, J. R. & Valette, J. P. (1971), *The Role of Trasnportation in Regional Economic Development* (Lexington Mass.: D. C. Heath).

Kraft, J. & Kraft, A. (1979), 'Benefits and costs of low rent public housing', *Journal of Regional Science*, vol. 19, no. 3.

Krutilla, J. V. (1967), 'Conservation reconsidered', *American Economic Review*, vol. 57, no. 2.

Krutilla, J. V. & Eckstein, O. (1958), *Multiple Purpose River Development* (Baltimore: Johns Hopkins Press).

Lave, J. R. & Lave, L. B. (1970), 'Hospital cost functions', *American Economic Review*, vol. 60, no. 3.

Lave, L. B. & Seskin, E. P. (1970), 'Air pollution and human health', *Science*, vol. 169, no. 3947.

Lave, L. B. & Seskin, E. P. (1977), *Air Pollution and Human Health* (Baltimore: Johns Hopkins Press).

Layard, R. (1972), 'Introduction', in R. Layard (ed.), *Cost–Benefit Analysis* (Harmondsworth, Penguin).

Lean, W. (1967), 'Economic studies and assessment of town development', *Journal of the Town Planning Institute*, vol. 53, no. 4.

LeFevre, A. J. & Pickering, J. F. (1972), 'The economics of moving Covent Garden market', *Journal of Agricultural Economics*, vol. 23, no. 1.

Lever, W. F. (1974), 'Regional multipliers and demand leakages at establishment level', *Scottish Journal of Political Economy*, vol. 21, no. 2.

Lever, W. F. (1975), 'Regional multipliers and demand leakages at establishment level: a reply', *Scottish Journal of Political Economy*, vol. 22, no. 1.

Lichfield, N. (1964), 'Cost–benefit analysis in plan evaluation', *Town Planning Review*, vol. 35, no. 2.

Lichfield, N. (1966a), *Cost–Benefit Analysis in Town Planning: A Case Study of Cambridge* (Cambridge: Cambridgeshire and Isle of Ely County Council).

Lichfield, N. (1966b), 'Cost–benefit analysis in town planning: a case study of Swanley', *Urban Studies*, vol. 3, no. 3.

Lichfield, N. (1967), 'The evaluation of capital investment projects in town centre redevelopment', *Public Administration*, vol. 45, no. 2.

Lichfield, N. (1968), 'Economics in town planning', *Town Planning Review*, vol. 39, no. 1.

Lichfield, N. (1969), 'Cost–benefit analysis in urban expansion: a case study – Peterborough', *Regional Studies*, vol. 3, no. 2.

Lichfield, N. (1970), 'Evaluation methodology of urban and regional plans: a review', *Regional Studies*, vol. 4, no. 2.

Lichfield, N., & Chapman, H. (1968), 'Cost–benefit analysis and road proposals for a shopping centre', *Journal of Transport Economics and Policy*, vol. 2, no. 3.

Lichfield, N. & Chapman, H. (1970), 'Cost–benefit analysis in urban expansion: a case study, Ipswich', *Urban Studies*, vol. 7, no. 2.

Lichfield, N. & Proudlove, A. (1976), *Conservation and Traffic: A Case Study of York* (York: Ebor Press).

Lichfield, N. & Wendt, P. F. (1969), 'Six English new towns: a financial appraisal', *Town Planning Review*, vol. 40, no. 3.

Lichfield, N., Kettle, P. & Whitbread, M. (1979), *Evaluation in the Planning Process* (London: Pergamon).

Lipsey, R. G. & Lancaster, K. (1956), 'A general theory of second best', *Review of Economic Studies*, vol. 24, no. 1.

Little, I. M. D. (1951), *A Critique of Welfare Economics* (Oxford: Clarendon).

Little, I. M. D. & Mirlees, J. A. (1974), *Project Appraisal and Planning for Developing Countries* (London: Heinemann).

Little, I. M. D. & Scott, M. F. G. (1976), *Using Shadow Prices* (London: Heinemann).

Maass, A. (1966), 'Benefit–cost analysis: its relevance to public investment decisions', *Quarterly Journal of Economics*, vol. 80, no. 2.

McBride, G. A. (1970), 'Policy matters in investment decision-making', *Regional Studies*, vol. 4, no. 2.

McConnell, K. E. (1977), 'Congestion and willingness to pay: a study of beach use', *Land Economics*, vol. 53, no. 2.

McDowall, S. (1975), 'Regional multipliers and demand leakages at establishment level: a comment', *Scottish Journal of Political Economy*, vol. 22, no. 1.

McGuire, A. (1983), 'The regional income and employment impacts of nuclear power stations', *Scottish Journal of Political Economy*, vol. 30, no. 3.

McGuire, M. C. & Garn, H. A. (1969), 'The integration of equity and efficiency criteria in public project selection', *Economic Journal*, vol. 79, no. 316.

McKean, R. N. (1968), 'The use of shadow prices', in S. B. Chase (ed.), *Problems in Public Expenditure Analysis* (Washington, DC: Brookings Institution).

McMenamin, D. G. & Haring, J. E. (1974), 'An appraisal of non-survey techniques for estimating regional input–output models', *Journal of Regional Science*, vol. 14, no. 2.

McNicholl, I. H. (1981), 'Estimating regional industry multipliers', *Town Planning Review*, vol. 52, no. 1.

Majid, I., Sinden, J. A. & Randall, A. (1983), 'Benefit evaluation of increments to existing systems of public facilities', *Land Economics*, vol. 59, no. 4.

Manning, S. & Viscek, D. (1977), 'Measuring the economic impact of a community college system', *Annals of Regional Science*, vol. 11, no. 3.

Mansfield, N. W. (1971), 'The estimation of benefits from recreation sites and the provision of a new recreational facility', *Regional Studies*, vol. 5, no. 2.

Mao, J. C. T. (1966), 'Efficiency in public urban renewal expenditures through benefit–cost analysis', *Journal of the American Institute of Planners*, vol. 32, no. 2.

Marglin, S. A. (1962), 'Objectives of water resource development: a general statement', in A. Maass *et al.* (eds.), *Design of Water Resource Systems* (Cambridge, Mass.: Harvard University Press).

Marglin, S. A. (1963a), 'The social rate of discount and the optimal rate of investment', *Quarterly Journal of Economics*, vol. 77, no. 1.

Marglin, S. A. (1963b), 'The opportunity costs of public investment', *Quarterly Journal of Economics*, vol. 77, no. 2.

Marin, A. (1983), 'Your money or your life', *Three Banks Review*, no. 138.

Marquand, J. (1980), 'Measuring the effects and costs of regional incentives', Paper no. 32, UK Department of Industry.

Masser, I. (1972), *Analytical Methods for Urban and Regional Planning* (Newton Abbot: David & Charles).

Mathur, V. K. & Rosen, H. S. (1974), 'Regional employment multiplier: a new approach', *Land Economics*, vol. 50, no. 1.

Mazumdar, D. (1974), 'The rural–urban wage gap, migration and the shadow wage', World Bank Staff Working Paper no. 197, Washington, DC.

Merrett, A. J. & Sykes, A. (1966), *Capital Budgeting and Company Finance* (London: Longmans).

Messner, S. D. (1968), 'Urban redevelopment in Indianapolis: a benefit–cost analysis', *Regional Studies*, vol. 8, no. 2.

Miernyk, W. H. (1965), *The Elements of Input–Output Analysis* (New York: Random House).

Miernyk, W. H. (1976), 'Comments on recent developments in regional input–output analysis', *International Regional Science Review*, vol. 1, no 2.

Miller, R. E. (1969), 'Inter-regional feedbacks in input–output models: some empirical results', *Western Economic Journal*, vol. 7, no. 1.

Mishan, E. J. (1970), 'What is wrong with Roskill?', *Journal of Transport Economics and Policy*, vol. 4, no. 3.

Mishan, E. J. (1974), 'Flexibility and consistency in project evaluation', *Economica*, vol. 41, no. 161.

Mishan, E. J. (1976), 'The uses of compensating and equivalent variations in cost-benefit analysis', *Economica*, vol. 43, no. 170.

Mishan, E. J. (1982a), *Cost–Benefit Analysis*, 3rd edn (London: Allen & Unwin).

Mishan, E. J. (1982b), 'The new controversy about the rationale of economic evaluation', *Journal of Economic Issues*, vol. 16, no. 1.

Mohring, H. (1961), 'Land values and the measurement of highway benefits', *Journal of Political Economy*, vol. 69, no. 3.

Moncur, J. T. (1975), 'Estimating the value of alternative outdoor recreation facilities within a small area', *Journal of Leisure Research*, vol. 7, no. 4.

Mooney, G. H. (1978), 'Planning for balance of care for the elderly', *Scottish Journal of Political Economy*, vol. 25, no. 2.

Moore, B. & Rhodes, J. (1973), 'Economic and exchequer implications of regional policy', Expenditure Committee (Trade and Industry Sub-Committee), Minutes of Evidence, Session 1972–73, HMSO.

Moore, B. & Rhodes, J. (1974), 'The effects of regional policy in the United Kingdom', in M. E. C. Sant (ed.), *Regional Policy and Planning for Europe* (Aldershot: Gower).

Moore, B. & Rhodes, J. (1976), 'A quantitative analysis of the effects of the regional employment premium and other regional policy instruments', in A. Whiting (ed.), *The Economics of Industrial Subsidies* (London: HMSO).

Moore, B. & Rhodes, J. (1977), *Methods of Measuring the Effects of Regional Policies* (Paris: OECD).

Moore, C. L. & Sufrin, S. C. (1974), 'The impact of a nonprofit institution on regional income', *Growth and Change*, vol. 5, no. 1.

Moore, F. T. & Peterson, J. P. (1955), 'Regional analysis: an inter-industry model of Utah', *Review of Economics and Statistics*, vol. 37, no. 4.

Munro, J. M. (1969), 'Planning the Appalachian development highway systems: some critical questions', *Land Economics*, vol. 45, no. 2.

Murray, M. P. (1975), 'The distribution of tenant benefits in public housing', *Econometrica*, vol. 43, no. 4.

Musgrave, R. A. (1969), 'Cost–benefit analysis and the theory of public finance', *Journal of Economic Literature*, vol. 7, no. 3.

Mushkin, S. J. (1962), 'Health as an investment', *Journal of Political Economy*, vol. 70, no. 5.

National Economic Development Council (NEDC) (1963), *Conditions Favourable to Economic Growth* (London: HMSO).

Needleman, L. (1969), 'The comparative economics of improvement and new building', *Urban Studies*, vol. 6, no. 2.

Needleman, L. (1980), 'The valuation of changes in risk of death by those at risk', *Manchester School of Economic and Social Studies*, vol. 48, no. 3.

Needleman, L. & Scott, B. (1964), 'Regional problems and location of industry policy in Britain', *Urban Studies*, vol. 1, no. 2.

Neenan, W. B. (1971), 'Distribution and efficiency in benefit–cost analysis',

Canadian Journal of Economics, vol. 4, no. 2.

Nelson, J. & Tweeten, L. (1973), 'Subsidized labour mobility: an alternative use of development funds', *Annals of Regional Science*, vol. 7, no. 1.

Nelson, J. P. (1982), 'Highway noise and property values: a survey of recent evidence', *Journal of Transport Economics and Policy*, vol. 16, no. 2.

Nicol, W. R. (1982), 'Estimating the effects of regional policy: a critique of the European experience', *Regional Studies*, vol. 16, no. 3.

Nwaneri, V. C. (1970), 'Equity in cost–benefit analysis: a case study of the Third London Airport', *Journal of Transport Economics and Policy*, vol. 4, no. 3.

Olsen, E. O. & Barton, D. M. (1983), 'The benefits and costs of public housing in New York City', *Journal of Public Economics*, vol. 20, no. 2.

Organization for Economic Cooperation and Development (OECD) (1969), *Manual of Industrial Project Analysis*, Vols. 1 and 2 (Paris: OECD).

P.A. Management Consultants (1972), *Cost–Benefit Analysis in Social Services for the City of Leicester* (London: P.A. Management Consultants Ltd).

Peaker, A. (1976), 'New primary roads and sub-regional economic growth: further results: a comment on J. S. Dodgson's paper', *Regional Studies*, vol. 10, no. 1.

Pearce, D. W. (1971), *Cost–Benefit Analysis* (London: Macmillan).

Pearce, D. W. (1983), *Cost–Benefit Analysis*, 2nd edn (London: Macmillan).

Pearce, D. W. & Nash, A. J. (1973), 'The evaluation of urban motorway schemes: a case study – Southampton', *Urban Studies*, vol. 10, no. 2.

Pearce, D. W. & Nash, A. J. (1981), *The Social Appraisal of Projects* (London: Macmillan).

Pearse, P. H. (1968), 'A new approach to the evaluation of non-priced recreational resources', *Land Economics*, vol. 44, no. 1.

Pearse, P. H. (1972), 'A new approach to the evaluation of non-priced recreational resources: a rejoinder', *Land Economics*, vol. 48, no. 4.

Piachaud, D. & Weddell, J. M. (1972), 'The economics of treating varicose veins', *International Journal of Epidemiology*, vol. 1, no. 2.

Pigou, A. C. (1932), *The Economics of Welfare*, 4th edn (London: Macmillan).

Porter, R. C. (1979), 'Secondary markets in benefit–cost analysis', mimeo, University of Michigan.

Poulton, M. C. (1982), 'A land use evaluation technique for decision-makers', *Regional Studies*, vol. 16, no. 2.

Prest, A. R. & Turvey, R. (1965), 'Cost–benefit analysis: a survey', *Economic Journal*, vol. 75, no. 4.

Province of British Columbia (1977), *Guidelines for Benefit–Cost Analysis* (Victoria: Environment and Land Use Committee Secretariat).

Quarmby, D. A. (1967), 'Choice of travel mode for the journey to work', *Journal of Transport Economics and Policy*, vol. 1, no. 3.

Richardson, H. W. (1972), *Input–Output and Regional Economics* (London: Weidenfeld & Nicolson).

Ridker, R. G. & Henning, J. A. (1967), 'The determinants of residential property values with special reference to air pollution', *Review of Economics and Statistics*, vol. 49, no. 2.

Rinehart, J. R. (1963), 'Rates of return on municipal subsidies', *Southern Economic Journal*, vol. 29, no. 3.

Roberts, M. (1974), *An Introduction to Town Planning Techniques* (London: Hutchinson).

Rothenberg, J. (1965), 'Urban renewal programs', in R. Dorfman (ed.), *Measuring Benefits of Government Investments* (Washington, DC: Brookings Institution).

Rothenberg, J. (1967), *Economic Evaluation of Urban Renewal* (Washington, DC: Brookings Institution).

Sant, M. (1975), *Industrial Movement and Regional Development* (Oxford: Pergamon).

Sassone, P. G. & Shaffer, W. A. (1978), *Cost–Benefit Analysis* (New York: Academic Press).

Sazama, G. W. (1970), 'A benefit–cost analysis of a regional development incentive: state loans', *Journal of Regional Science*, vol. 10, no. 3.

Schaffer, W. H. & Chu, K. (1969), 'Non-survey techniques for constructing regional inter-industry models', *Papers, Regional Science Association*, vol. 23.

Schofield, J. A. (1976a), 'The economic return to preventive social work', *International Journal of Social Economics*, vol. 3, no. 3.

Schofield, J. A. (1976b), 'Economic efficiency and regional policy in Britain', *Urban Studies*, vol. 13, no. 2.

Schofield, J. A. (1978), 'Some evidence on the economic return to DREE's industrial development activity', *Canadian Journal of Regional Science*, vol. 1, no. 1.

Schulze, W. D., d'Arge, R. C. & Brookshire, D. S. (1981), 'Valuing environmental commodities: some recent evidence', *Land Economics*, vol. 57, no. 2.

Schwind, P. J. (1977), 'The evaluation of land use alternatives: a case study of the metropolitan fringe of Honolulu, Hawaii', *Land Economics*, vol. 53, no. 4.

Scitovsky, T. (1941), 'A note on welfare propositions in economics', *Review of Economic Studies*, vol. 9, no. 1.

Scott, A. (1977), 'The test rate of discount and changes in the base-level income in the United Kingdom', *Economic Journal*, vol. 86, no. 2.

Scott, M. F. G., MacArthur, J. D. & Newbery, D. M. G. (1976), *Project Appraisal in Practice* (London: Heinemann).

Scottish Development Department (1968), *The Central Borders: A Plan for Expansion* (Edinburgh: HMSO).

Seeley, I. H. (1973), *Outdoor Recreation and the Urban Environment* (London: Macmillan).

Self, P. (1975), *Econocrats and the Policy Process* (London: Macmillan).

Sen, A. K. (1967), 'Isolation, assurance and the social rate of discount', *Quarterly Journal of Economics*, vol. 81, no. 1.

Shaffer, R. & Tweeten, L. (1974), 'Measuring net economic changes from rural industrial development: Oklahoma', *Land Economics*, vol. 50, no. 3.

Shoup, D. C. (1973), 'Cost effectiveness of urban traffic law enforcement', *Journal of Transport Economics and Policy*, vol. 7, no. 1.

Silberberg, E. (1972), 'Duality and the many consumers' surpluses', *American Economic Review*, vol. 62, no. 4.

Silberberg, E. (1978), *The Structure of Economics* (New York: McGraw Hill).

Sinclair, C. (1969), 'Costing the hazards of technology', *New Scientist*, vol. 44, no. 671.

Sinclair, M. T. & Sutcliffe, C. M. S. (1978), 'The Keynesian regional income multiplier', *Scottish Journal of Political Economy*, vol. 25, no. 2.

Sinclair, M. T. & Sutcliffe, C. M. S. (1982), 'Keynesian income multipliers with first and second round effects: an application to tourist expenditure', *Oxford Bulletin of Economics and Statistics*, vol. 44, no. 4.

Sinclair, W. F. (1976), *The Economic and Social Impact of the Kemano II Hydro-Electric Project on British Columbia's Fisheries Resources Value* (Vancouver:

Fisheries and Marine Service, Environment Canada).

Sjaastad, L. A. & Wisecarver, D. L. (1977), 'The social cost of public finance', *Journal of Political Economy*, vol. 85, no. 3.

Small, K. A. (1977), 'Estimating the air pollution costs of transport modes', *Journal of Transport Economics and Policy*, vol. 11, no. 2.

Smith, J. (1982), 'Threshold analysis of two potential metropolitan growth corridors: a Melbourne case study', *Town Planning Review*, vol. 53, no. 2.

Smith, R. J. (1971), 'The evaluation of recreation benefits: the Clawson method in practice', *Urban Studies*, vol. 8, no. 2.

Smith, R. J. (1975), 'Problems of interpreting recreation benefits from a recreation demand curve', in G. A. C. Searle (ed.), *Recreational Economics and Analysis* (London: Longmans).

Squire, L. & van der Tak, H. G. (1975), *Economic Analysis of Projects* (Baltimore: Johns Hopkins Press).

Starkie, D. N. M. & Johnson, D. M. (1975), *The Economic Value of Peace and Quiet* (Farnborough: Saxon House).

Steele, D. B. (1969), 'Regional multipliers in Great Britain', *Oxford Economic Papers*, vol. 21, no. 2.

Steele, D. B. (1972), 'A numbers game (or the return of regional multipliers)', *Regional Studies*, vol. 6, no. 2.

Stevens, B. H. & Trainer, G. H. (1976), 'The generation of error in regional input–output impact models', RSRI working paper A1-76, Amherst, Mass.

Stokey, E. & Zeckhauser, R. (1978), *A Primer for Policy Analysis* (New York: Norton).

Stone, P. A. (1963), *Housing, Town Development Land and Costs* (London: Estates Gazette).

Stopher, P. R. & Meyburg, A. H. (1976), *Transportation Systems Evaluation* (Lexington, Mass.: D. C. Heath).

Straszheim, M. R. (1972), 'Researching the role of transportation in regional development', *Land Economics*, vol. 48, no. 3.

Sugden, R. (1972), 'Cost-benefit analysis and the withdrawal of railway services', *Bulletin of Economic Research*, vol. 24, no. 1.

Sugden, R. & Williams, A. (1978), *The Principles of Practical Cost–Benefit Analysis* (Oxford: Oxford University Press).

Sumka, H. J. & Stegman, M. A. (1978), 'An economic analysis of public housing in small cities', *Journal of Regional Science*, vol. 18, no. 3.

Sumner, M. T. (1980), 'Benefit–cost analysis in Canadian practice', *Canadian Public Policy*, vol. 6, no. 2.

Swales, J. K. (1975), 'Regional multipliers and demand leakages at establishment level: a comment', *Scottish Journal of Political Economy*, vol. 22, no. 1.

Swan, N. & Glynn, A. (1977), 'The costs and benefits of industrial location grants', Discussion Paper no. 93, Economic Council of Canada.

Swaney, J. A. & Ward, F. A. (1985), 'Optimally locating a national public facility: an empirical application of consumer surplus theory', *Economic Geography*, vol. 61, no. 2.

Thaler, R. H. & Rosen, S. (1975), 'The value of saving a life: evidence from the labour market', in E. E. Terleckyj (ed.), *Household Production and Consumption* (New York: Columbia University Press).

Thomas, S. (1978), 'The valuation of accident cost savings', *Journal of Transport Economics and Policy*, vol. 12, no. 3.

Thompson, J. M. (1967), 'An evaluation of two proposals for traffic restraint in central London', *Journal of the Royal Statistical Society*, Series A, vol. 130, no. 3.

Thurow, L. C. & Rappoport, C. (1969), 'Law enforcement and cost–benefit analysis', *Public Finance*, vol. 24, no. 1.

Tinbergen, J. (1957), 'The appraisal of road construction: two calculation schemes', *Review of Economics and Statistics*, vol. 39, no. 3.

United Nations Industrial Development Organization (UNIDO) (1972), *Guidelines for Project Evaluation* (New York: United Nations).

US Water Resources Council (1970), *Standards for Planning Water and Land Resources* (Washington, DC: WRC).

US Water Resources Council (1973), 'Water and related land usages: establishment of principles and standards for planning', *Federal Register*, no. 38.

Varian, H. R. (1979), 'Notes on cost–benefit analysis', mimeo, University of Michigan.

Vickerman, R. W. (1974), 'The evaluation of benefits from recreational projects', *Urban Studies*, vol. 11, no. 3.

Vickerman, R. W. (1975), *The Economics of Leisure and Recreation* (London: Macmillan).

Wabe, J. S. (1971), 'A study of house prices as a means of establishing the value of journey time, the rate of time preference and the valuation of some aspects of environment in the London Metropolitan region', *Applied Economics*, vol. 3, no. 3.

Wager, R. (1972), *Care of the Elderly – An Exercise in Cost–Benefit Analysis* (London: Institute of Municipal Treasurers and Accountants).

Walden, M. L. (1981), 'A note on benefit and cost estimates in publicly assisted housing', *Journal of Regional Science*, vol. 21, no. 3.

Walter, G. W. & Schofield, J. A. (1977), 'Recreation management: a programming example', *Land Economics*, vol. 53, no. 2.

Walters, A. A. (1975), *Noise and Prices* (London: Clarendon).

Warford, J. J. & Williams, A. (1971), 'Rural water supplies and the economic evaluation of alternative location patterns', in M. G. Kendall (ed.), *Cost–Benefit Analysis* (London: English Universities Press).

Watson, P. L. (1974), *The Value of Time: Behavioural Models of Modal Choice* (Lexington, Mass.: D. C. Heath).

Watson, P. L. & Mansfield, N. (1973), 'The valuation of time in cost–benefit studies', in J. N. Wolfe (ed.), *Cost–Benefit and Cost Effectiveness* (London: Allen & Unwin).

Weingartner, H. M. (1963), *Mathematical Programming and the Analysis of Capital Budgeting Problems* (New York: Prentice-Hall).

Weisbrod, B. A. (1961), *Economics of Public Health* (Philadelphia: University of Pennsylvania Press).

Weisbrod, B. A. (1962), 'Education and investment in human capital', *Journal of Political Economy*, vol. 70, no. 5.

Weisbrod, B. A. (1968), 'Income redistribution effects and benefit–cost analysis', in S. B. Chase (ed.), *Problems in Public Expenditure Analysis* (Washington, DC: Brookings Institution).

Weisbrod, B. A. (1971), 'Costs and benefits of medical research: a case study of poliomyelitis', *Journal of Political Economy*, vol. 79, no. 3.

Weisbrod, B. A. (1981), 'Benefit–cost analysis of a controlled experiment: treating the mentally ill', *Journal of Human Resources*, vol. 16, no. 4.

Weiss, S. J. & Gooding, E. C. (1968), 'Estimation of differential employment multipliers', *Land Economics*, vol. 44, no. 2.

Wildavsky, A. (1966), 'The political economy of efficiency: cost–benefit analysis, systems analysis and program budgeting', *Public Administration Review*, vol. 26, no. 4.

Williams, A. (1973), 'CBA: bastard science and/or insidious poison in the body politik?' in J. N. Wolfe (ed.), *Cost Benefit and Cost Effectiveness* (London: Allen & Unwin).

Williams, A. & Anderson, R. W. (1975), *Efficiency in the Social Services* (Oxford: Martin Robertson).

Willig, R. D. (1976), 'Consumer's surplus without apology', *American Economic Review*, vol. 66, no. 4.

Willis, K. G. (1980), *The Economics of Town and Country Planning* (London: Granada).

Willis, K. G. (1982), 'Green belts: an economic appraisal of physical planning policy', *Planning Outlook*, vol. 25, no. 2.

Willis, K. G. (1985), 'Estimating the benefits of job creation from local investment subsidies', *Urban Studies*, vol. 22, no. 2.

Willis, K. G. & Whisker, P. M. (1980), 'Economic assessment of local authority aid to industry', *Planning Outlook*, vol. 23, no. 2.

Willis, K. G., Curtis, S. & Giziakis, E. (1979), 'Observation and assessment centres for children', *International Journal of Social Economics*, vol. 6, no. 3.

Wilson, J. H. (1975), 'The impact of a non-profit institution on regional income: a discussion', *Growth and Change*, vol. 6, no. 3.

Wilson, J. H. (1977), 'Impact analysis and multiplier specification', *Growth and Change*, vol. 8, no. 3.

Wilson, J. H. & Raymond, R. (1973), 'The economic impact of a university upon the local community', *Annals of Regional Science*, vol. 7, no. 2.

Wilson, R. A. (1980), 'Rate of return to becoming a qualified scientist or engineer in Great Britain, 1966–76', *Scottish Journal of Political Economy*, vol. 27, no. 1.

Wilson, T. (1968), 'The regional multiplier', *Oxford Economic Papers*, vol. 20, no. 3.

Worswick, G. D. N. (1972), 'Is progress in economic science possible?', *Economic Journal*, vol. 82 no. 325.

Wright, K. G. (1974), 'Alternative measures of the output of social programmes: the elderly', in A. J. Culyer (ed.), *Economic Policies and Social Goals* (London: Martin Robertson).

Yannopoulos, G. (1973), 'Local income effects of office relocation', *Regional Studies*, vol. 7, no. 1.

Index